CONTROL AND DYNAMIC SYSTEMS

Advances in Theory and Applications

Volume 45

CONTRIBUTORS TO THIS VOLUME

JAMES S. ALBUS

DONALD S. BLOMQUIST

HOWARD P. BLOOM

GARY P. CARVER

RICHARD H. F. JACKSON

ALBERT JONES

PHILIP NANZETTA

JOHN A. SIMPSON

DENNIS A. SWYT

CONTROL AND DYNAMIC SYSTEMS

ADVANCES IN THEORY AND APPLICATIONS

Volume Editor

RICHARD H. F. JACKSON

Manufacturing Engineering Laboratory
National Institute of Standards & Technology
Gaithersburg, Maryland

Edited by

C. T. LEONDES

School of Engineering and Applied Science
University of California, Los Angeles
Los Angeles, California
and
College of Engineering
University of Washington
Seattle, Washington

VOLUME 45: MANUFACTURING AND
AUTOMATION SYSTEMS:
TECHNIQUES AND TECHNOLOGIES
Part 1 of 5
Three Pillars of Manufacturing Technology

ACADEMIC PRESS, INC.
Harcourt Brace Jovanovich, Publishers
San Diego   New York   Boston
London   Sydney   Tokyo   Toronto

ACADEMIC PRESS RAPID MANUSCRIPT REPRODUCTION

Academic Press, Inc.
1250 Sixth Avenue, San Diego, California 92101

United Kingdom Edition published by
Academic Press Limited
24–28 Oval Road, London NW1 7DX

Library of Congress Catalog Number: 64-8027

International Standard Book Number: 0-12-012745-8

PRINTED IN THE UNITED STATES OF AMERICA
92 93 94 95 96 97   BC   9 8 7 6 5 4 3 2 1

CONTENTS

CONTRIBUTORS

Numbers in parentheses indicate the pages on which the authors' contributions begin.

James S. Albus (197), *Robot Systems Division, Manufacturing Engineering Laboratory, National Institute of Standards and Technology, Gaithersburg, Maryland 20899*

Donald S. Blomquist (163), *Automated Production Technology Division, Manufacturing Engineering Laboratory, National Institute of Standards and Technology, Gaithersburg, Maryland 20899*

Howard P. Bloom (31), *Factory Automation Systems Division, Manufacturing Engineering Laboratory, National Institute of Standards and Technology, Gaithersburg, Maryland 20899*

Gary P. Carver (31), *Factory Automation Systems Division, Manufacturing Engineering Laboratory, National Institute of Standards and Technology, Gaithersburg, Maryland 20899*

Richard H. F. Jackson (1), *Manufacturing Engineering Laboratory, National Institute of Standards and Technology, Gaithersburg, Maryland 20899*

Albert Jones (249), *Manufacturing Engineering Laboratory, National Institute of Standards and Technology, Gaithersburg, Maryland 20899*

Philip Nanzetta (307), *Office of Manufacturing Programs, Manufacturing Engineering Laboratory, National Institute of Standards and Technology, Gaithersburg, Maryland 20899*

John A. Simpson (17, 333), *Manufacturing Engineering Laboratory, National Institute of Standards and Technology, Gaithersburg, Maryland 20899*

Dennis A. Swyt (111), *Precision Engineering Division, National Institute of Standards and Technology, Gaithersburg, Maryland 20899*

PREFACE

At the start of this century, national economies on the international scene were, to a large extent, agriculturally based. This was, perhaps, the dominant reason for the protraction, on the international scene, of the Great Depression, which began with the Wall Street stock market crash of October 1929. In any event, after World War II the trend away from agriculturally based economies and toward industrially based economies continued and strengthened. Indeed, today, in the United States, approximately only 1% of the population is involved in the agriculture industry. Yet, this small segment largely provides for the agriculture requirements of the United States, and, in fact, provides significant agriculture exports. This, of course, is made possible by the greatly improved techniques and technologies utilized in the agriculture industry.

The trend toward industrially based economies after World War II was, in turn, followed by a trend toward service-based economies; and, in fact, in the United States today roughly 70% of the employment is involved with service industries, and this percentage continues to increase. Nevertheless, of course, manufacturing retains its historic importance in the economy of the United States and in other economies, and in the United States the manufacturing industries account for the lion's share of exports and imports. Just as in the case of the agriculture industries, more is continually expected from a constantly shrinking percentage of the population. Also, just as in the case of the agriculture industries, this can only be possible through the utilization of constantly improving techniques and technologies in the manufacturing industries in what is now popularly referred to as the second Industrial Revolution. As a result, this is a particularly appropriate time to treat the issue of manufacturing and automation systems in this international series. Thus, this is Part 1 of a five-part set of volumes devoted to the most timely theme of "Manufacturing and Automation Systems: Techniques and Technologies."

"Three Pillars of Manufacturing Technology" is the title of this volume. It is edited by Richard H. F. Jackson, and its coauthors are Dr. Jackson and his colleagues at the National Institute of Standards and Technology (NIST) Manufacturing Engineering Laboratory, a unique organization on the international scene.

Ordinarily, in this Academic Press series, the series editor provides the overview of the contents of the respective volumes in the Preface. However, in the case of this unique volume, this, among other things, is provided by Dr. Jackson in the first chapter of this volume. Therefore, suffice it to say here that Dr. Jackson and his colleagues are all to be most highly commended for their efforts in producing a most substantively important volume that will be of great and lasting significance on the international scene.

C. T. Leondes

ACKNOWLEDGMENT

There are many who contributed to the production of this book whose names do not appear on title pages. Principal among these are Barbara Horner and Donna Rusyniak, and their participation bears special mention. Their editorial and organizational skills were invaluable in this effort, and their untiring and cheerful manner throughout made it all possible.

Richard H. F. Jackson

MANUFACTURING TECHNOLOGY: THE THREE PILLARS

RICHARD H.F. JACKSON

Manufacturing Engineering Laboratory
National Institute of Standards and Technology
Gaithersburg, MD 20899

I. INTRODUCTION

In their simplest form, the three pillars of manufacturing technology are: robots, computers, and production equipment (or, in the case of mechanical manufacturing: machine tools). Since the mid-twentieth century and the onslaught of the second industrial revolution, these three pillars have formed the foundation upon which all new techniques and advances in the technology of factory floor systems have been built. This is of course not to deny the importance of advances in non-technology areas such as lean production, quality management, continuous improvement, and workforce training. On the contrary, improvements in these areas can provide significant gains in industrial productivity and competitiveness. Nevertheless, in the area of manufacturing systems the three pillars are just that: the foundation. Because of their importance to the manufacturing systems of today, and because they will also be critical to the development of the advanced manufacturing systems of tomorrow, they are the theme of this volume. Further, these pillars form the foundation of the advanced manufacturing research at the Manufacturing Engineering Laboratory (MEL) of the National Institute of Standards and Technology (NIST), and since that work is at the center of the U.S. government's programs in advanced manufacturing research and development, it is featured here[1]. The chapters were produced by staff of MEL and can be considered a profile of a successful research and development effort in manufacturing technology, which profile is aimed at providing guidance for those who would expand on it.

II. A VISION OF MANUFACTURING IN THE TWENTY-FIRST CENTURY

To thrive in the twenty-first century, a manufacturing enterprise must be globally competitive, produce the highest quality product, be a low cost, efficient producer, and be responsive to the market and to the customer. In short, the next century's successful manufacturing firms will be "World Class Manufacturers" who make world class products for world class customers; i.e, customers who know precisely what they want and are satisfied only with world-class products.

There are many perspectives from which one can view a world class manufacturing firm. It can be viewed from the perspective of the shop floor and its interplay of the hardware and software of production. It can be viewed from the perspective of the legal environment in which it operates, both nationally and internationally. It can be viewed from the standpoint of the business environment, with its complex of tax and capital formation policies that affect long- and short-term planning. It can be viewed from the perspective of its corporate structure and embedded management techniques, which may facilitate or impede progress toward a manufacturing system of the twenty-first century. It may be viewed through the eyes of its employees, and the way it incorporates their ideas for improving the manufacturing process, the way it seeks to train them and upgrade their skills, and the way it strives to integrate them with the intelligent machines of production. It may be viewed from the perspective of its policies for performing, understanding, incorporating and transferring state-of-the-art research in advanced manufacturing technology.

A world class manufacturer may be viewed from these and many other perspectives, but, as depicted in Figure 1, the essential issue of importance for a successful twenty-first century manufacturing enterprise is to learn how to operate smoothly and efficiently within each of these regimes and to combine them into a smoothly functioning, well-oiled engine of production. Such an engine takes as input the classical categories of labor, capital, and material, and produces world class products for world class customers.

---

[1]See the Appendix to this Introduction for a brief description of the Manufacturing Engineering Laboratory and its programs.

21ST CENTURY MANUFACTURING

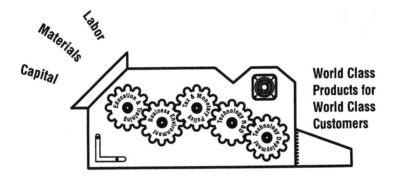

Figure 1. Twenty-First Century Manufacturing.

While each of these perspectives is certainly important in its own right, and the interplay with the other factors is critical, in this book we concentrate on two of the gears of the well-oiled machine: manufacturing technology research and development, and technology deployment. We concentrate on these because these are the only appropriate areas for a scientific and engineering laboratory like NIST to address. In fact, NIST is the only federal research laboratory with a specific mission to support U.S. industry by providing the measurements, calibrations, quality assurance techniques, and generic technology required by U.S. industry to support commerce and technological progress.

III. THE THREE PILLARS OF MANUFACTURING TECHNOLOGY

We have organized this book and our programs at NIST around the three basic components of any successful mechanical manufacturing system: machine tools and basic precision metrology, intelligent machines and robotics, and computing and information technology, all overlaid on a background of standards and measurement technology, as shown in Figure 2. These are the three pillars of manufacturing technology research and development because they have been, are, and will be for some time to come, the quintessential components of factory floor systems from the very small to the very large. That is, in one fashion or another they have been addressed on factory floors since the establishment of the second manufacturing paradigm [1]. Further, in one fashion or another, they will continue to be addressed by those who seek to refine this paradigm over the next twenty to fifty years. Measurement is important because, simply put, if you cannot measure, you cannot manufacture to specifications. Standards are important to this foundation because, among other things, they help define the interfaces that allow interconnections among the elements of manufacturing enterprises, and the subsequent interchange of process and product information.

THREE PILLARS OF MANUFACTURING TECHNOLOGY RESEARCH AND DEVELOPMENT

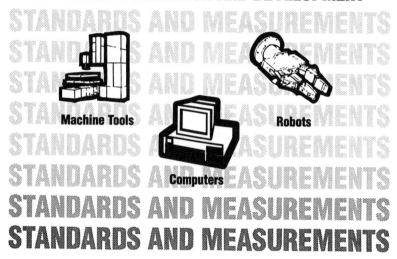

Figure 2. Three Pillars of Manufacturing Technology R&D.

It is not the intent of this introductory chapter to dwell in detail on these three pillars: that is accomplished in the remaining chapters. Nevertheless, it is important to provide some additional detail here, so as to put in context those remaining chapters. For example, in the area of computing and information technology, a twenty-first century manufacturer can be depicted as shown in Figure 3. This figure and the information technology aspects of it are discussed fully in Chapter 3 of this book. The figure depicts the enormous amount of information that will be available in such an enterprise, and the importance of providing a smoothly functioning, seamless mechanism for making all the information available to whomever needs it, whenever and wherever it may be needed. These enterprises may be collocated as indicated on the left or they may be distributed throughout the country, or indeed the world. In any case, these information-laden enterprises must find ways to collect, store, visualize and access the information that is required for process-monitoring and decision-making. The exploded view at the right shows how one portion of such an enterprise would be integrated through the sharing of process and product data. The development of product data exchange standards is a critical component in the development of such an integrated enterprise.

INFORMATION TECHNOLOGY

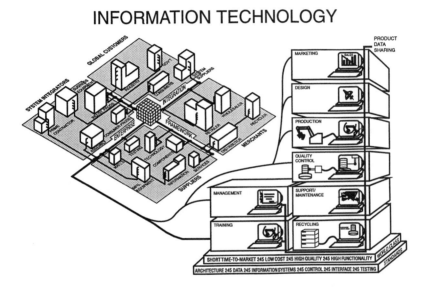

Figure 3. Information Technology in the Twenty-First Century

Indeed, the thesis is put forth in Chapter 3 that the development of product data exchange technology and international agreement on standards for the exchange of such data form the last real opportunity for U.S. industry to catch up to, and perhaps even leapfrog, the Japanese manufacturing juggernaut. Some in this country believe that there is very little about Japanese manufacturing techniques that we do not know. The challenge for us is to find a way to apply those techniques within our distinctly U.S. culture of creativity and entrepreneurial drive of the individual and the independent enterprise, and to apply it both in the advancement of technology and in the conduct of business. Product data exchange standards and the technologies like concurrent engineering and enterprise integration that are subsequently enabled by them may just be the answer to this challenge. For example, students of manufacturing technology have long held that one of the keystones of Japanese manufacturing success is the "Keiretsu," a large, vertically integrated corporation with complete control of the manufacturing process from raw material to distribution, achieved through stable corporate structures and person-to-person interactions. Complete integration of all aspects of a manufacturing enterprise from design to manufacturing to marketing and sales, in a homogeneous, collocated or widely diverse and distributed enterprise is, in a sense, an "electronic Keiretsu." Such electronic Keiretsu could provide for U.S. manufacturers the "level playing field" required to compete successfully against the vertically integrated traditional Keiretsu in Japan. Product data exchange standards are the heart and soul of concurrent engineering and enterprise integration, and are critical to the ability of U.S. Manufacturers to survive into the next century. More on this and on the emerging technologies of concurrent engineering and enterprise integration is given in Chapter 3.

In the area of machine tools and basic metrology, the next pillar, impressive gains have been made in recent years, and will continue to be made in the future. The machine tools of the next century will exploit these gains and will be capable of heretofore unheard-of precision, down to the nanometer level. These gains will be possible through the combination of new techniques in machine tool quality control and in our ability to see, understand, and measure at the atomic level. After all, the truism that in order to build structures, one must be able to measure those structures, applies at the atomic level also. Even today, there are U.S. automobile engines manufactured with tolerances of 1.5 micrometers, pushing the limits of current machine tool technology. Next generation machine tool technology is depicted in Figure 4. It is noteworthy that both machine tools and coordinate measuring machines will make gains in precision based on the same kind of technologies, and thus it will be possible to have both functions available in one collection of cast iron. The metrological aspects of achieving such precision are further discussed in Chapter 4. The process control issues are discussed further in Chapter 5.

Intelligent machines today and tomorrow go far beyond the simple industrial robots and automated transporters of yesterday. These new machines are a finely tuned combination of hardware components driven by a software controller with enough built-in intelligence as to make it almost autonomous. These software controllers will be capable of accumulating information from an array of advanced sensors and probing devices, matching that information against their own world-model of the environment in which they operate, and computing optimum strategies for accomplishing their tasks. These strategies must be determined in real time and take into account the dynamic environment in which these machines will operate. The sensory information obtained must of course conform to existing standards of manufacturing data exchange. This can only be accomplished

MANUFACTURING AND MEASURING TO SUPER-PRECISION

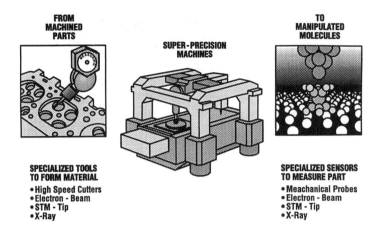

Figure 4. Machine Tools in the Twenty-First Century.

through a full understanding of the nature of intelligence and the development of software tools and structures which facilitate their efficient implementation and realization in a software controller. The architecture for such a controller is illustrated in Figure 5, which indicates the hierarchical nature of the controller and the processes for interaction with sensors and world models. NIST work in this area, as well as in the development of a true theory of intelligence, is discussed in Chapter 6 of this book.

The core of the NIST effort in advanced manufacturing is the Automated Manufacturing Research Facility (AMRF), and no discussion of NIST work in advanced manufacturing would be complete without some discussion of it. The AMRF has played a significant role in the identification and development of new and emerging technologies, measurement techniques, and standards in manufacturing. In fact, it was a catalyst in the legislative process that resulted in the Technology Competitiveness Act of 1988, which changed the name of the National Bureau of Standards to NIST and enhanced our mission. Further, the successful work described in the other chapters would not have been possible without the development of this research facility. The AMRF is a unique engineering laboratory. The facility provides a basic array of manufacturing

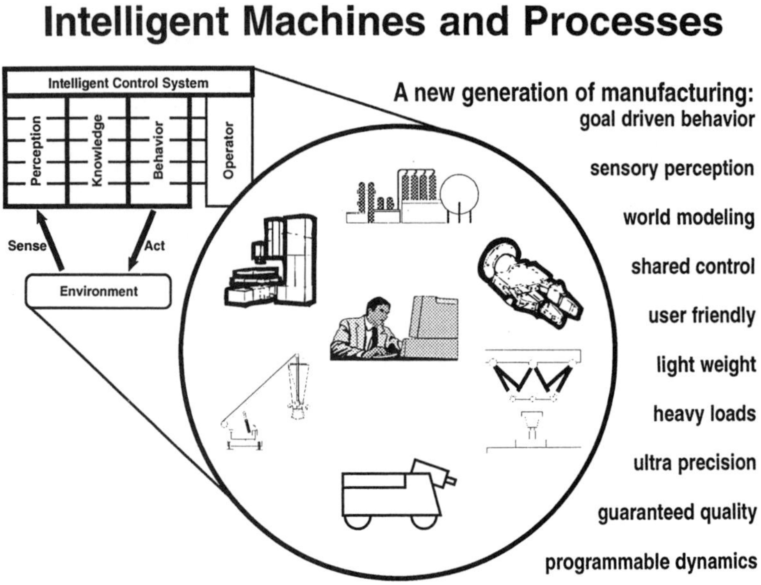

Intelligent Machines and Processes

Figure 5. Intelligent Machines in the Twenty-First Century.

equipment and systems, a "test bed," that researchers from NIST, industrial firms, universities, and other government agencies use to experiment with new standards and to study new methods of measurement and quality control for automated factories. In this sense, it serves as kind of "melting pot" of ideas from researchers across a broad spectrum of disciplines, talents, and skills. Credit should be given here to the U.S. Navy's Manufacturing Technology Program for helping to support the AMRF financially as well as in identifying fruitful opportunities for research.

The AMRF includes several types of modern automated machine tools, such as numerically controlled milling machines and lathes, automated materials-handling equipment (to move parts, tools, and raw materials from one "workstation" to another), and a variety of industrial robots to tend the machine tools. The entire facility operates under computer control using an advanced control approach, based on the ideas in Figure 5, and pioneered at NIST. The architecture is discussed further in Chapter 6. The AMRF incorporates some of the most advanced, most flexible automated manufacturing techniques in the world. The key to a fully flexible, data-driven, automated manufacturing system is the control system. Its architecture must be capable of integrating the separate architectures of production management, information management, and communications management. The evolution of some of these architectures is the subject of chapter 7 of this book.

No amount of successful scientific or engineering research will have an effect on our nation's competitiveness unless these ideas are successfully transferred to, and deployed by, U.S. industry. It is NIST's mission to do so, and our history shows extraordinary success in this area. Nevertheless, there is always room for improvement, and the enhanced mission ascribed to NIST in 1988, includes some valuable new programs in this outreach area. Chapter 8 of this book describes some of the NIST work in this area of technology transfer and deployment, from early times to more recent times.

Lastly, chapter 9 contains a more fully fleshed-out discussion of the twenty-first century manufacturing system. It is based on observations of the history of manufacturing, a recognition of the changing paradigms in this nationally critical area, and a full understanding of the importance of the three pillars of manufacturing technology research and development.

IV. HISTORICAL CONTEXT AND OUTLOOK

The United States is at a critical juncture in its manufacturing history. Since the first industrial revolution, the U.S. manufacturing sector has maintained a position of strength in competition for world market share. The strength of this industrial base has provided incredible growth in the U.S. gross national product and contributed immensely to the material well-being of the citizenry. Unfortunately, this position and its beneficial effects on the standard of living can no longer be taken for granted. U.S. industries are being threatened from all sides: market share has been slipping, capital equipment is becoming outdated, and the basic structure of the once mighty U.S. corporation is being questioned.

A growing national debate has focused on this decline of U.S. industry's competitiveness and the resultant loss of market share in the global marketplace. This rapid loss of competitiveness of American industry in international markets is an extremely serious problem with wide-ranging consequences for the United States' material well-being, political influence, and security. The national debate on this subject has identified many possible culprits, ranging from trade deficits to short-term, bottom-line thinking on the part of U.S. management. Nevertheless, among them certainly are the slow rate at which new technology is incorporated in commercial products and processes, and the lack of attention paid to manufacturing. There is a clear need to compete in world markets with high-value-added products, incorporating the latest innovations, manufactured in short runs with flexible manufacturing methods. Research, management, and manufacturing methods that support change and innovation are key ingredients needed to enhance our nation's competitive position. In fact, efforts in these areas seem to be paying off already. In his upbeat message on technology opportunities for America [2], John Lyons, the NIST Director, reports impressive gains recently in the cost of labor, productivity and the balance of trade.

As Lyons notes, one key area in which we must focus continued effort is in commercializing new technologies. As a nation, we have been slow to capitalize on new technology developed from America's own intellectual capability. Many ideas originating in the American scientific and technical community are being commercially exploited in other parts of the world. In the past, small and mid-sized companies have led U.S. industry in innovation. The nation must now find ways to help such companies meet the demands of global competition, when the speed with which firms are able to translate innovations into quality commercial products and processes is of utmost importance.

Success in this effort will only come from full cooperation among government, industry, and academe. Theis was clearly stated by the President's Commission on Industrial Competitiveness [3]:

"Government must take the lead in those areas where its resources and responsibilities can be best applied. Our public leaders and policy must encourage dialogue and consensus-building among leaders in industry, labor, Government, and academia whose expertise and cooperation are needed to improve our competitive position.... Universities, industry, and Government must work together to improve both the quality and quantity of manufacturing related education."

Many agendas have been written for how such cooperation should proceed and what issues must be addressed by each of these sectors. (See, for example [4-16]). This book describes some efforts underway at NIST to aid U.S. manufacturers in their own efforts to compete in the global marketplace, and thrive in the next century. At the center of these efforts are the programs and projects of the Manufacturing Engineering Laboratory at the National Institute of Standards and Technology. Thus, the chapters focus on these programs and projects.

The first chapter after this introduction provides some historical background on the importance of measurement and standards in manufacturing. Following that are four chapters discussed above that in essence are set pieces on NIST work in the three pillars of manufacturing technology. These are followed by the chapter on architecture development in our Automated Manufacturing Research Facility, a chapter on our programs in technology transfer, and a closing chapter on the future of manufacturing.

REFERENCES:

1. J.A. Simpson, "Mechanical Measurement and Manufacturing," in "Three Pillars of Manufacturing Technology," R.H.F. Jackson, ed., Academic Press, 1991.

2. J.W. Lyons, "America's Technology Opportunities," "The Bridge," National Academy of Engineering, Vol. 21, No. 2, Summer 1991.

3. Presidential Commission on Industrial Competitiveness, January 1985.

4. "U.S. Technology Policy," Executive Office of the President, Office of Science and Technology Policy, Washington, DC 20506, September 26, 1990.

5. "The Challenge to Manufacturing: A Proposal for a National Forum," Office of Administration, Finance, and Public Awareness, National Academy of Engineering, 2101 Constitution Avenue, NW, Washington, DC, 1988.

6. "Making Things Better: Competing in Manufacturing," U.S. Congress, Office of Technology Assessment, OTA-ITE-443, U.S. Government Printing Office, Washington, DC 20402-9325, February 1990.

7. "The Role of Manufacturing Technology in Trade Adjustment Strategies," Committee on Manufacturing Technology in Trade Adjustment Assistance, Manufacturing Studies Board, National Research Council, National Academy Press, 2101 Constitution Avenue, NW, Washington, DC 20418, 1986.

8. "Toward a New Era in U.S. Manufacturing - The Need for a National Vision," Manufacturing Studies Board, Commission on Engineering and Technical Systems, National Research Council, National Academy Press, 2101 Constitution Avenue, NW, Washington, DC 20418, 1986.

9. "Bolstering Defense Industrial Competitiveness - Preserving our Heritage the Industrial Base Securing our Future," Report to the Secretary of Defense by the Under Secretary of Defense (Acquisition), July 1988.

10. "Paying the Bill: Manufacturing and America's Trade Deficit," U.S. Congress, Office of Technology Assessment, OTA-ITE-390, U.S. Government Printing Office, Washington, DC 20402, June 1988.

11. M.R. Kelley and H. Brooks, "The State of Computerized Automation in U.S. Manufacturing," Center for Business and Government, October 1988. (Order from: Weil Hall, John F. Kennedy School of Government, Harvard University, 79 John F. Kennedy Street, Cambridge, MA 02138.)

12. "A Research Agenda for CIM, Information Technology," Panel on Technical Barriers to Computer Integration of Manufacturing, Manufacturing Studies Board and Cross-Disciplinary Engineering Research Committee jointly Commission on Engineering and Technical Systems, National Research Council, National Academy Press, 2101 Constitution Avenue, NW, Washington, DC 20418, 1988.

13. I.C. Magaziner and M. Patinkin, "The Silent War - Inside the Global Business Battles Shaping America's Future," Random House, Inc., New York, 1989.

14. C.J. Grayson, Jr. and C. O'Dell, "American Business - A Two-Minute Warning - Ten changes managers must make to survive into the 21st century," The Free Press, A Division of Macmillan, Inc., 866 Third Avenue, New York, NY 10022, 1988.

15. S.S. Cohen and J. Zysman, "Manufacturing Matters - The Myth of the Post-Industrial Economy," Basic Books, Inc., New York, 1987.

16. M. L. Dertouzos, R.K. Lester, R.M. Solow, and the MIT Commission on Industrial Productivity, "Made in America - Regaining the Productive Edge," The MIT Press, Cambridge, MA, 1989.

APPENDIX: THE MANUFACTURING ENGINEERING LABORATORY AT NIST

Since the Manufacturing Engineering Laboratory and its programs are featured in this volume, we provide in this Appendix a brief discussion of the mission, goals, and objectives of the Laboratory and its five Divisions.

The Manufacturing Engineering Laboratory (MEL) is one of eight technical laboratories that comprise the National Institute of Standards and Technology. Laboratory staff maintain competence in, and develop, technical data, findings, and standards in manufacturing engineering, precision engineering, mechanical metrology, mechanical engineering, automation and control technology, industrial engineering, and robotics to support the mechanical manufacturing industries. The laboratory develops measurement methods particularly suited to the automated manufacturing environment as well as traditional mechanical measurement services. The Laboratory also develops the technical basis for interface standards between the various components of fully automated mechanical manufacturing systems. The Laboratory consists of a staff of approximately 300, with an annual budget of approximately $30 million, both of which are devoted to the program goal, "to contribute to the technology base which supports innovation and productivity enhancement in the mechanical manufacturing industries on which the Nation's economic health depends."

The Laboratory is managed by the Director, the Deputy Director, the Executive Officer, the Manager of the Office of Manufacturing Programs, the Manager of the Office of Industrial Relations, and the Chiefs of the five Divisions: the Precision Engineering Division, the Automated Production Technology Division, the Robot Systems Division, the Factory Automation Systems Division, and the Fabrication Technology Division. Each of these Division's programs, is discussed in more detail below.

The Precision Engineering Division:

This five-group division develops and conveys metrological principles and practices to support precision-engineered systems vital to U.S. manufacturing industries. Its general work encompasses the physics and engineering of systems which generate, measure and control to high resolution the length-dimensional quantities of position, distance, displacement and extension. Specific work focusses on: electron-, optical- and mechanical-probe microscopies; coordinate measuring machines; and metrology-intensive production systems for automated manufacturing, precision metal-working and integrated-circuit fabrication. Of special concern are industries dealing with the fabrication of nanometer-scale structures such as micro-machined mechanical components and lithographically produced electronic devices. This is an emerging technology certain to be of

major commercial significance, involving fierce international competition, and
requiring broad advances in metrology for its support.

The Robot Systems Division:

The Robot Systems Division is composed of five groups: the Intelligent Control
Group, the Sensory Intelligence Group, the Systems Integration Group, the
Unmanned Systems Group, and the Performance Measures Group. The goal of
the Robot Systems Division is to develop the technology base necessary to
characterize and measure the performance of, and establish standards for, future
generations of intelligent machines systems. To do so will require the extension
and fusion of the concepts of artificial intelligence and the techniques of modern
control to make control systems intelligent and make artificial intelligence systems
operate in real-time.

The control system architecture developed for the Automated Manufacturing
Research Facility (AMRF) at the National Institute of Standards and Technology
had its foundations in robotics research. The concepts of hierarchical control,
task decomposition, world modeling, sensory processing, state evaluation, and
real-time sensory-interactive, goal-directed behavior are all a part of this control
architecture. These concepts are derived from the neurosciences, computer
science, and control system engineering. The result of the AMRF project is a
generic modular intelligent control system with well defined interfaces. This
control architecture provides a first generation bridge between the high level
planning and reasoning concepts of artificial intelligence and the low level sensing
and control methods of servo mechanisms.

Industrial robots and computer controlled machine tools for manufacturing are
only the first of many potential applications of robotics. Future applications will
be in less structured, more unpredictable, and often more hostile environments.
The application of robotics and artificial intelligence to other sectors of industry,
commerce, space, and defense has the potential to vastly increase productivity,
improve quality, and reduce the cost of many different kinds of goods and
services.

The Robot Systems Division expects that the robotics technology currently
available or being developed (mostly with NASA and Department of Defense
funding) will be applied to produce a wide variety of intelligent machine systems.
The Division proposes to position itself to meet the measurement and standards
needs of this emerging industry by being actively involved in a variety of
developments so that it understands the underlying technology and discerns the
measurement and standards requirements.

The Robot Systems Division is currently working on applications of the AMRF
control system architecture to space telerobots for satellite servicing, to coal mine
automation, shipbuilding, construction, large scale (high payload, long reach)
robots, multiple cooperating semi-autonomous land vehicles (teleoperated and
supervised autonomy), and nuclear submarine operational automation systems.

The Factory Automation Systems Division:

The Factory Automation Systems Division develops and maintains expertise in information technology and engineering systems for design and automated manufacturing. This Division is composed of two offices and four groups: the CALS/PDES Project Office, the IGES/PDES/STEP Office, the Product Data Engineering Group, the Production Management Systems Group, the Integrated Systems Group, and the Machine Intelligence Group. They provide leadership in the development of national and international standards relating to information technology for design and product data exchange to enhance the quality, performance, and reliability of automated systems for manufacturing. In addition, they perform research and develop the technology for integrated database requirements to support product life cycles, requirements that can be reflected in both standards for product data exchange and for information technology frameworks needed to manage the data. They also perform research on methodologies such as concurrent engineering as a means of promoting the production of improved quality products for U.S. industry, and in information technology in such areas as distributed database management, information modeling, and computer networks. They develop validation, verification, and testing procedures for emerging product data exchange standards and for the performance of information technology systems for manufacturing, and perform research in the information technology requirements for such manufacturing processes as product design, process planning, equipment control, inspection, and logistics support.

The Automated Production Technology Division:

The Automated Production Technology Division is composed of six Groups: the Ultrasonic Standards Group, the Acoustic Measurements Group, the Sensor Systems Group, the Sensor Integrations Group, the Mass Group, and the Force Group. The Division develops and maintains competence in the integration of machine tools and robots up to and including the manufacturing cell level. The Division develops and maintains computer-assisted techniques for the generation of computer codes necessary for the integration of machine tools and robots. The Division also develops the interfaces and networks necessary to combine robots and machine tools into workstations and workstations into manufacturing cells. In addition, the Division performs the research and integration tests necessary for in-process monitoring and gaging. The Staff maintains competence in engineering measurements and sensors for static and dynamic force-related quantities and other parameters required by the mechanical manufacturing industry. They also conduct fundamental research on the nature of the measurement process and sensory interaction, as well as the development, characterization, and calibration of transducers used in discrete parts manufacturing.

The Fabrication Technology Division:

This Division has three Groups: the Support Activities Group, the Special Shops Group, and the Main Shops Group. The Fabrication Technology Division designs, fabricates, repairs, and modifies precision apparatus, instrumentation, components thereof, and specimens necessary to the experimental research and development work of NIST. Services include: engineering design, scientific instrument fabrication, numerically controlled machining, welding, sheet metal fabrication, micro fabrication, grinding, optical fabrication, glassblowing, precision digital measuring, tool crib, and metals storeroom. The Division also develops and maintains competence in CAD/CAM, automated process planning, and shop management systems.

MECHANICAL MEASUREMENT AND MANUFACTURING

JOHN A. SIMPSON

Manufacturing Engineering Laboratory
National Institute of Standards and Technology
Gaithersburg, MD 20899

I. INTRODUCTION

The history of mechanical measurement and manufacturing are completely intertwined. Each new development metrology gave rise to new manufacturing procedures. Each new manufacturing development required new levels of measurement precision and accuracy. In a broad outline, the history of the interaction between measurement and manufacturing from the viewpoint of a single manufacturer is detailed in the study by Ramchandran Jaikumar [1] of the Beretta Arms Company.

Another viewpoint where this interplay is vividly displayed is in the modern relationships between the development of modern manufacturing and the programs of the National Bureau of Standards (NBS), predecessor of the National Institute of Standards and Technology (NIST). The author has been responsible for the NBS Dimensional Metrology Program since 1968, and for the Program in Manufacturing Engineering since 1978.

II. THE BIRTH OF MANUFACTURING

Prior to approximately 1800, it is impossible to speak of the manufacturing of mechanical parts. They were <u>crafted</u>, each one expressing the skill of the mechanic that made them. There were few, if any, measurements. When dimensions were needed, they were obtained by simple caliper comparison from an original sample made by the master mechanic.

A. The English System

Manufacturing, in any real sense, began in England at the Woolwich Arsenal in 1789. Here, Henry Maudslay combined previous recent inventions made by others, such as the lathe slide rest, the lead screw driven by change gears and by 1800 had produced the engine lathe for rotational parts, which remains virtually unchanged today. The same technology was soon extended to milling machinery. For the first time, it was possible to create precision measuring tools, such as screw micrometers, and then use them to control the dimensional accuracy of the parts being made. The process fed upon itself; each improvement in machine permitted more precise instruments; each improvement in instruments lead to better machines. Measurements to a precision of a few thousandths of an inch were soon possible by craftsmen on the factory floor.

B. The Beginning of Drafting

Prior to 1800, Mechanical Drawings were unknown. The only, what we might call, "manufacturing data structures" were physical models, one for each part and a set for each craftsman. The second great manufacturing invention of the turn of the 19th century occurred in France, when Gaspard Monge in 1801 published La Geometrie Descriptive. This remarkable work not only defined the now standard three-view, third-quadrant orthographic projection, but also for the first time added dimensions to the drawing. He created a "manufacturing data structure" infinitely reproducible, still ubiquitous today. Again, the use of dimensions led to more measurements and the instruments to perform them.

   With the machine tool and micrometer of Maudslay and the mechanical drawing of Monge, an enabling technology of mechanical measurement and manufacturing was complete. For the next one hundred and fifty years, it was to form the enabling technology of the First Manufacturing Paradigm.

C. The American System

The next major step in the development of manufacturing and measurement took place in the United States. In 1789, Eli Whitney was carrying out a contract with the Army to produce muskets. He was using the pre-manufacturing technology, with one significant difference. He was using jigs and fixtures to allow the employment of less skilled labor, and in the process invented the "go-no-go" gaging system. In its simplest form, this system to control the size of a hole, for example, had a small plug which must "go" into the hole and a larger plug which must not fit, "no-go," into the hole. Gaging has the advantage of being much faster and easier than measurement but does not result in a number and hence is of somewhat limited usefulness. Although now subject to controversy as to its effectiveness, most screw threads are still gaged by "go", "no-go" technology.

The use of gages moved measurement off the factory floor to the metrology laboratory where the "calibration" of the gages by measurement was performed.

By 1855, Samuel Colt, in Hartford, Connecticut, had combined the gaging system of Whitney and the machine tools of Woolwich, and the First Manufacturing Paradigm was complete.

III. THE PERFECTING OF A MANUFACTURING PARADIGM, 1860 - 1920

The Colt paradigm was adopted widely, both here and abroad, and the next sixty years were spent in creating improved machine tools, gages and measuring equipment. However, the concept was unchanged and the changes were minor. The major changes occurred in Workshop Management. The major creator of the new system was Frederick Taylor who, under the banner of "Scientific Management," introduced the practice of breaking down all tasks to their smallest component so that they could be handled by virtually unskilled labor. Also, by use of "time studies", stop-watch measurements, his followers determined the length of time required for each operation. Taylor attempted to define the one-best, least time-consuming, way of accomplishing each task and had management insist that they be done this way and no other.

At the same time, what is known as "hard automation", was developed. Similar in spirit to Taylorism, this technology developed machines which automatically performed, with maximum efficiency, one relatively simple operation on one or a very small family of parts. A change in product required a virtual rebuilding of the machine. The specialization, however, resulted in very high rates of production, and "hard automation" still reigns supreme for lowest unit cost. Measurement continued to be by special purpose gages unique to each product and dimension. This production system led to Henry Ford's mass production triumphs.

During this time, measurement technology improved from what it was in 1850 only in detail. The only significant change was the invention and development between 1890 and 1907 by Carl Edvard Johansson of Sweden of the gage-block [2]. These artifacts could be "wrung" together to form stacks of almost arbitrary length and could serve as a surrogate for any gage with parallel surfaces. This invention added greatly to the flexibility of gaging, reducing the number of gages needed and adding to precision of the gaging process. Gage blocks remain the working standards for length to this day.

IV. THE PARADIGM IS COMPLETE 1950, - SCIENCE AND MATHEMATICS ENTER MANUFACTURING AND MEASUREMENT

Neither manufacturing nor measurement, both having arisen in the workshop, has ever had much of a formal theoretical basis. Except for some work by Ernst Abbe in the 1890's, measurement has none. The manufacturing literature, if we discard those texts dealing with workshop management, has little more.

In the late 1920's, W. Shewhart [3] of Western Electric turned his attention to the possibility of, in a mass-production environment of reducing the amount of measurement effort, by the use of statistical sampling of the output. By 1939, Shewhart had developed most of the tools of statistical quality control, such as control charts and their use, and was giving graduate courses in the procedures. His teaching, was further developed during the 1940's by W. Edwards Demming [4], and others.

The needs of efficient mass production of munitions in World War II resulted in the wide adoption of the technology. Demming later introduced the statistical technology to Japan in the 1950's, where it is generally considered to be, in large measure, responsible for the current Japanese reputation for quality products.

To understand this development, one must remember that the essence of the American System, as opposed to the philosophy of Maudslay, which was to strive for "perfection," was to make a product that was just adequate for its intended use; that is, to reduce cost and increase speed of production to allow the widest possible tolerances on every dimension. The Statistical Quality Control philosophy is an extension of this concept and seeks to determine the minimum number of measurements you must make to assure, on a large population of supposedly identical parts, that only a pre-selected percentage would, on average, lie outside a given tolerance band. The increase in productivity gained in such a system is large, and it assures a level of average quality. However, it knowingly allows a given percentage of the delivered product to be non-conforming.

In Statistical Process Control, the measurement data taken on the product for quality control is used to characterize the process and can be used to determine trends in that process, and after some time delay, allow action to be taken to stop the drift into excessive non-conformity.

These techniques formed the basis for Quality Control in the latter days of the first manufacturing paradigm.

C. Eisenhart and W.J. Youden, working at The National Bureau of Standards, were the first to realize that "measurement itself is a production process whose product is numbers" and, hence, these same techniques are applicable to metrology, the science of measurement [5].

This idea of measurement as a production process, with quality monitored by statistical means, was brilliantly realized by P. Pontius, and J. Cameron [6] in the NBS calibration program. Initially applied to the calibration of mass standards, it soon was extended to all calibrations where there was sufficient volume performed to constitute a suitable statistical population. For the first time, NBS could provide uncertainty statements that were scientifically based. By means of Measurement Assurance Programs, (MAP), these accurate uncertainty statements could be propagated down the standards chain to the workshop. This system has become the standard procedure throughout the entire measurement community and is often called for in procurement contracts.

During the same period, the philosophers of science, notably R. Carnap [7], turned their attention to the theory of measurement. Carnap propounded an axiometric theory of what constitutes a "proper measurement" and the means necessary that a measurement need never be repeated since any competent metrologist will get the same results within the stated precision of the initial result. Such a measurement, he called "proper". This theory applied to calibration activities greatly influenced those activities.

The actual measurement processes remained essentially what they were in 1850, direct comparison between the test object and an artifact "standard". Now, however, the comparison was often done by electronic instruments. One still needed an artifact standard for each and every object to be calibrated or measured. The 1850 Colt Paradigm was now complete.

V. A SECOND PARADIGM EMERGES, 1950-1970

While the finishing touches were being applied to the first paradigm, a new one was being born. In the late 1940's, the U.S. Air Force Sponsored work at MIT to develop programmable machine tools. The basic idea could be compared to applying to metal cutting the technique of the Jacquard loom for making patterned fabric. The machine's sequence of operations were pre-programed on paper tape, and the operator's function reduced to loading, unloading, tool changing, and looking for problems. Although the first commercial programmable, or Numerical Control (NC) machine tool, came a decade later, major adoption of the technology had to wait another decade for the development of improved electronics. The computer developments soon led to Computer Numerical Control (CNC), which eliminated the paper tape, and Direct Numerical Control (DNC), which put a group of machines under the control of a central computer.

The technology was revolutionary. It meant that an element of flexibility was reintroduced into metal part production, which had been missing since before 1850. To change the product, one merely had to re-program, not re-build the machine. This increased flexibility is at the heart of the second paradigm and turned out to produce profound changes in manufacturing practice.

A similar change was appearing in measurement technology. The computer eased the task of Statistical Process Control, but in addition, Coordinate Measuring Machines (CMM) were beginning to appear in advanced manufacturing and measurement facilities. These machines, originally based on Jig-grinders, had the capability of making three dimensional, accurate measurements of any object within a three dimensional volume limited only by machine size. Although the machines were slow, and required a skilled operator, they broke the Colt paradigm of the need for an artifact standard for every product. Now both fabrication and measurement were flexible.

VI. THE NATIONAL BUREAU OF STANDARDS JOINS THE REVOLUTION

The dimensional metrology program of NBS in 1968 represented the Colt paradigm in its finest flower. It was equipped with hundreds of artifact standards, highly developed mechanical comparators, and a highly skilled, experienced and devoted staff of calibration technologists with years of experience in precise comparison measurement. Having recently moved to the new Gaithersburg, MD, site, the laboratories were new and had state-of-the-art environmental controls, but all was not well. The calibration staff was being decimated by retirements, and suitable replacements were difficult to find. A change in Government policy required that the cost of calibrations be recovered by fees which were high and rising rapidly.

E. Ambler, then Director of the Institute of Basic Standards, later Director of NBS, who was responsible for the Metrology Programs of NBS, concluded that the solution to the problem was to automate the calibration process. What expertise in automation and electronics there was at the Bureau at that time, lay mostly in the Atomic Physics Program. Within the Metrology Division, even electronics had not been adopted, in the age of the flowering of the transistor, there were only a few dozen vacuum tubes and no transistors in the entire laboratory. Ambler, therefore, combined the two and assigned members of the Atomic Physics Program the task of automating the dimensional metrology laboratory.

To those of us involved, it seemed that the best solution was to cross breed the measuring machine with the newly affordable minicomputer and create a computer controlled coordinate measuring machine that could be programmed in a high level language and perform, at least, the more taxing calibrations on this new machine. With the aid of a budget initiative from Congress, a measuring machine was procured from Moore Special Tool of Bridgeport Connecticut, and a PDP 7 mini-computer from Digital Products. A five-man team, consisting of two computer experts and three mechanical engineers, was hired to staff the effort.

Just as the measuring machine program started, NBS came under pressure from the gear manufacturers to expand the program in gear metrology. Under the existing, then dominant, Colt paradigm this would have meant obtaining dozens of artifact standard gears to cover the desired range of type and size.

In an act of considerable political courage, and even more optimism, the management announced that not only were we not extending our range of master gears, but were shutting down the existing gear laboratory since from now on all gear calibrations would be done in our new measuring machine facility. The decision did not meet with universal acclaim.

The "die was cast". Although we would, and do, continue to support many industries relying on artifact standards, all our research and development efforts in dimensional metrology from that time on have been directed at development of the second paradigm.

VII. THE SECOND PARADIGM DEVELOPS, 1975-1980

By the mid 1970's, the Bureau had a Computer Controlled Measurement Machine that was programmable in the BASIC programming language and capable of completely unmanned operation. It made possible studies that had been previously impossible because of the amount of data to be acquired.

Moreover, since Laser Interferometers had become available, it was now possible to completely map the twenty one degrees of freedom error matrix of the three-axis measuring machine to high accuracy. We discovered that the machine was 5 to 20 times more reproducible in its behavior than it was accurate. This suggested that if software within the computer controller was written to correct for these in-built errors, appreciable increase in accuracy could be realized. This improvement soon proved to be the case.

The technique was soon extended to real-time correction for temperature and temperatures gradients. It is interesting that over a decade had to pass before the first commercial CMM, using software accuracy enhancement, came onto the market to great success [8].

The same ability to take masses of data soon led R.Hocken [9] and collaborators to apply a principle originally developed for photogrammetry: the use of "Super Redundant" algorithms. Such algorithms make use of redundant measurements on an object, not only for the classic role of determining uncertainty of the measurements, but also for characterization of the machine in the process of making measurements. For example, in the course of measuring a two-dimensional grid plate, one can simultaneously measure the angle between the X and Y axis of the machine and the relation between the X and Y scales. At the expense of considerable computational burden, the technique can be extended to all the twenty one error parameters of a three-axis machine.

The success of accuracy enhancement on CMMs suggested that even greater gains in productivity could be obtained if the same technique were applied to metal cutting lathes and mills [10]. Early experiments showed that the same or greater accuracy enhancement was easily possible.

After proof of concept, NBS took on the task of producing stamp perforation cylinders for the Bureau of Printing and Engraving. This task required the drilling of several thousand 1mm holes in a pattern, corresponding to the shape of the stamp, on the surface of a 500cm diameter drum. Since the hole drum had to match a pin drum, the task required a non cumulative error of .025mm (0.0001in) in the two dimensional location of the holes.

Since the commercial machining center controller, unlike the NBS designed controller of the CMM, did not support software error correction, a scheme of passing the outputs of the carriage encoders through a computer, before entering the encoder readings into the controller, was devised. The computer, using lookup tables, corrected the readings for inbuilt errors and compensated for the temperature of the machine length scales, and passed the corrected reading to the machine controller, which was unaware of the substitution. This technique has continued to be valuable in other applications.

For the first time, the drum accuracy was sufficient that interchangeability between pin and hole cylinders was obtained. In the process of this development, another problem arose, drill breakage. Although the drills were replaced long before their expected life, on the average a drill broke every thousand holes. In a cylinder of fifty thousand holes, this was unacceptable. K.W. Yee and collaborators [11] developed "Drill Up," an acoustic sensor that measured the sound produced by the drilling operation and detected the change in sound as the drill dulled and ordered the drill replaced before it became dull enough to break. With this circuit in place, half a million holes were drilled without a single drill breakage. This device was commercialized.

The success of "Drill Up" and real-time temperature compensation gave rise to a more generalized concept of a "Self Conscious Machine" that was aware, by the use of measurements, of its state and took action when that state varied from the desired one. This concept is still developing.

The concept was further extended under the name of "Deterministic Metrology," sometimes called "Deterministic Manufacturing," by applying it to the total manufacturing process. In essence, the idea is that, within definable limits, mechanical manufacturing is a deterministic process and that every part attribute is controlled by one or more process parameters.

Hence, if a process ever produced one "good" part, it could not produce a "bad" part unless at least one process parameter had changed. If all process parameters were monitored by suitable measurement, when a change occurred, appropriate action could be taken before the first "bad" part was produced.

There are three conditions that must be fulfilled before such a system can be implemented.

1. The process must be completely automated without any humans directly involved in the processing. (Humans are not deterministic in behavior.)

2. The feedstock must be uniform or have been pre-characterized.

3. The process must be understood at the engineering level so that there exists a model connecting all part attributes to the controlling process parameters. It is notable that this is the same model that is necessary in Statistical Process Control (SPC), where the parameter change is deduced from changes in part attributes. In Deterministic Manufacturing, however, the model is now used in the opposite sense to deduce change in part attributes from changes in process parameters.

If such a manufacturing system can be implemented, the productivity gains are potentially very large. Now, all the measurements are made on process parameters rather than on part attributes, and there are far fewer process parameter measurements required than there are attribute measurements on each individual part, even if full use of statistical sampling is made. The individual measurements are more complex but with the aid of the micro-computer the complexity is easily dealt with.

Moreover, once a process is instrumented, the system is suitable for use on any part that the process is capable of making. A new product requires no new gaging system. Also, scrap is reduced almost to zero and no longer is it necessary to have up to one third of your workforce doing rework.

Perhaps the most important benefit lies in the fact that, unlike any of the statistical quality control schemes, one does not knowingly ship defective products. In the 1980's, customers were no longer satisfied with the typically two percent defects which was the industrial norm under the older paradigm.

If Deterministic Metrology is implemented, Manufacturing and Measurement become almost indistinguishable. No longer does one need metrology laboratories and the chain of calibrations leading from national standard artifacts to factory floor gages, only the relatively few sensors must be calibrated. The only real issue is, can such a scheme be, in fact, realized?

VIII. MEASUREMENT AND MANUFACTURING UNITE

After a NBS reorganization in 1978, which combined the Bureau program in Robotics and the Metrology program into the Center for Manufacturing Engineering and Process Technology (CMEPT), it was possible to attack the issue of realization of Deterministic Metrology.

The planning of an Automated Manufacturing Research Facility (AMRF), to satisfy the condition requiring a fully automated full scale manufacturing system on which to realistically test the idea, was begun [12].

Within the manufacturing community at this time, similar ideas were coming into realization. The first Flexible Manufacturing Systems (FMS), were being designed. These systems made use of Direct Numerical Controlled Machines with automated material handling equipment, often robots, loading the machines and passing material from one machine to the next. These FMS, somewhat resembling the earlier "hard automated" production lines, had the advantage that, within a part family, they could at least in principle produce a variety of parts with changes only in the computer program.

The speed with which these systems added value led to increased interest in "In-Process" gaging so that defects were caught earlier in the production cycle.

Increasingly, quality control was considered as part of production, and measurement increasingly moved out of the metrology laboratory onto the shop floor.

To supply these computer driven manufacturing systems with the required control code, the design process was also changing rapidly. Hand drafting was being replaced by Computer Aided Design (CAD) systems. The savings in time and effort were outstanding, but a whole new series of problems arose. These problems centered around interfaces. There were dozens of CAD systems on the market, each storing the manufacturing data in a different format. To design on one CAD system and to modify on another, required the production of a drawing and manually re-entering it into the new system.

The same interface problems existed between computers and the DNC systems they would control. Unlike the mechanical drawing, for which recognized standard practices have existed for decades, manufacturing data structures were in chaos.

The manufacturing community, led by the aerospace industries who were furthest advanced in the use of CAD, started a concerted attack on the CAD to CAD system interface problem. In 1979, funded by the Air Force, NBS set up a cooperative program with industry and standards bodies to create an Initial Graphics Exchange Specification (IGES), which would allow free interchange of mechanical drawings in digital form between CAD systems.

The strategy adopted was to develop a neutral data format, based on the various entities that make up a drawing, such as points, lines, curves and their intersections. These entities and their positions were stored in a strictly defined format in an ASCII file. Pre- and Post-Processors were to be written by the system vendors to transform from this format to and from the native format or language of the various systems. This strategy overcame one of the greatest barriers to standard interfacing: it protected the proprietary nature of the systems native languages or data formats. Under this scheme, the nature of these internal structures did not have to be revealed to the maker of the target system nor to any third party who would serve as a translator.

The strategy succeeded spectacularly. In less than two years, IGES was adopted as a National Standard (ANSI Y14.26M) and was soon universally adopted by the manufacturers of CAD systems. The initial Pre- and Post-Processors left much to be desired, but over time, they improved and new entities were added to the Standard. IGES, now in version 5, remains the principle interface standard for CAD to CAD exchange today, and a voluntary body continues to upgrade it. The interface problem of machine to machine and computer to machine, i.e., CAD to Computer Aided Manufacturing (CAM), remained. Within CMEPT/NBS, the decision was taken to use the opportunity to study both Deterministic Metrology and the interface problems at the same time on the AMRF.

Others at NBS and elsewhere were developing communication protocols for the communication links between factory floor elements. This work eventually led to the International Standards Organization/Open System Interface(ISO/OSI) communication protocols and the so-called GM MAP [13]. We decided to focus on the interpretation problems arising after the message had passed the element's operating system.

These decisions meant that the AMRF should have a control architecture which not only provided for the integration of the sensors needed for

Deterministic Metrology, but also for the integration of a heterogenous collection of different machine tools or robot controllers, and computers, each of which had a different native language (command set). It was also deemed necessary that the architecture be usable at any degree of automation and be capable of expansion, adding workstations, cells, or changing physical configuration without major software revision.

To accomplish these goals, use was made of a control architecture initially developed for robotics by J. Albus and coworkers [14]. The philosophy behind this system embodied some of the basic ideas of Artificial Intelligence and treats each element of the system as a finite state machine. Overall, the total system is a finite state machine whose elements, in turn, are such machines. As such, programming by "if then" statements or state tables, is straightforward, and the use of state graphs helps the visualization of the control sequences.

This architecture, which became known as the NBS Hierarchical Control Architecture, is highly modular in structure and makes use of common memory for all communication between machines or computer program modules.

This common memory or "post-box" system immediately gives the advantage of reducing the number of interfaces in a n element system from (n) times (n-1) to (n), since total interconnectivity is accomplished as soon as every element can communicate with the post-boxes.

If one combines the post-box idea with the IGES idea of the use of neutral data formats for the contents of those post-boxes with, one has a system that permits each element or even each computer program to have a different native language. The full implementation of this concept lead to the development of Manufacturing Data Management System, IMDAS, of considerable complexity and power [15].

Since each level of the hierarchy has the same structure, the human interface can be at any level, the human consisting of just another element with a computer screen and a key board whose command set is pre- and post-processed into the neutral data format as is any other. Hence, unlike most systems, transition from human control to fully automated via a mixed system presents no problem.

By 1981, with congressional initiative funding and support from the Navy Manufacturing Technology program, the construction of the AMRF in 5,000 square foot of floor space in the NBS Instrument Shops, was commenced. From the beginning, extensive cooperation from the private sector was enjoyed. Over the next few years, over 40 private companies sent Industrial Research Associates to work for months at a time with NBS on the project. Over twelve million dollars worth of equipment was donated or loaned for the effort.

The strategy was a success and, by 1988, a facility was in place, consisting of a horizontal machining center, a vertical machining center, a turning center, a cleaning and deburring center and a final inspection center, each with its robot and served by robot cart and automated storage and retrieval system. It was driven by an integrated control system, using the NBS Hierarchical Control

System, working from a digital representation, a predecessor of PDES, of the desired part [16].

The control architecture proved so advantageous that, besides its use in other processing applications, it was soon adopted under the name NASA Standard Reference Model (NASREM) as the standard model architecture for all NASA Space Station telerobotics. It is in use by the Department of Interior for underground mining robotic applications and has found military use for both undersea and land autonomous vehicles.

IX. THE SECOND MANUFACTURING PARADIGM MATURES

With the successful demonstration of the NBS AMRF, the development of control architectures and the success of various industrial factory floor experiments the "batch of one" has proven to be an economically viable manufacturing strategy. It appears that, for the time being, further advances in manufacturing productivity research will come not from focus on the shop floor, but from concentrating on the forty percent of the cost of a product incurred by operations performed before processing actually starts.

These operations include design, process planning, material resource planning, scheduling and optimization.

The current high interest in Simultaneous or Concurrent Engineering, where design and production planning are done simultaneously instead of sequentially, and Total Enterprise Integration and Computer Aided Acquisition and Logistical Support (CALS), attempts to move to a "paperless" environment for manufacturing, reflect the advanced manufacturing community's agreement with this shift in priorities.

Central to these efforts, is the world-wide effort to find a 21st Century replacement for the mechanical drawing as the dominant manufacturing data structure. The mechanical drawing has served well for almost 200 years but suffers a fatal flaw. Such a drawing is neither a complete nor unambiguous model of the described part. Much of the information needed to manufacture that part is contained only by text notes referencing other documents, such as standards for surface finish, heat treatments or tolerances. Moreover, since a drawing only shows edges and it is not unambiguous, it is possible to make "legal" drawings of parts which cannot exist in three-space.

Fortunately, humans, with training, can correctly "interpret" drawings in almost all cases. Computers, however, cannot "interpret", and it is impossible to write a process planning computer program to be driven by any representation of a mechanical drawing. Since IGES, at best, is a representation of a mechanical drawing, it shares this fatal flaw.

There is currently an international effort, under the International Standards Organization (ISO), to develop a "Standard for the Exchange of Product Model Data" (STEP), which will completely and unambiguously represent a part in digital form. In the U.S., there is a National program, "Product Data Exchange using STEP", involving hundreds of companies and government agencies to contribute to the definition of STEP.

X. NEW WORLDS TO CONQUER

In the immediate future, I believe, we will see the Second Manufacturing Paradigm become dominant. "Batch-of-One" manufacturing from highly flexible manufacturing systems will no longer be the rare exception. Computers operating from increasingly powerful data structures and data management algorithms will allow Enterprise Integration and Concurrent Engineering to become a reality, reducing the time-to-market. Time-to-market will increasingly become the arena in which the competitive battle will be fought. More slowly computer science and knowledge engineering will develop so that Knowledge Based Systems, containing machine intelligence, will become practical.

Measurement science will continue to improve. Both process and in-process metrology, in support of the Taguchi quality control idea of continuous improvement until production of scrap, is a rare event [17].

New manufacturing methods will continue to reduce the size of artifacts until both measurement and manufacture will be on the nanometer or atomic scale. Research headed in this direction is already underway.

We can be sure that after one hundred and forty years of measurement and manufacturing, the end of progress is not in sight. Details of work currently in hand at NIST are given in the succeeding chapters of this volume. The efforts of the entire manufacturing community will result in restructuring of manufacturing in the twenty first century. A **Vision of Manufacturing in the Twenty First Century** concludes this volume.

REFERENCES

1. R. Jaikumar, "From Filing and Fitting to Flexible Manufacturing: A Study of Process Control," Working Paper 88-045, Harvard Business School (1988).

2. W.R. Moore, "Foundations of Mechanical Accuracy," Moore Special Tool Co., p. 154-158 (1970).

3. W.A. Shewhart, "Economic Control of Quality of Manufactured Product," D. Van Nostrand Company, Inc., New York, NY (1931).

4. W.E. Demming, "Some Theory of Sampling," John Wiley & Sons, Inc., New York, NY (1950).

5.   C. Eisenhart, "Realistic Evaluation of the Precision and Accuracy of Instrument Calibration Systems," JOURNAL OF RESEARCH of the National Bureau of Standards, C. Engineering and Instrumentation 67c, No. 2 (1963).

6.   P.E. Pontius and Cameron, "Realistic Uncertainties and the Mass Measurement Process," National Bureau of Standards, Monograph 103 (1967).

7.   R. Carnap, "Philosophical Foundations of Physics," Basic Books, New York (1956).

8.   The Apollo series from Bendix Sheffield.

9.   R. Hocken, J.A. Simpson, et al, "Three Dimensional Metrology," CIRP Ann., 26, p. 403-408 (1977).

10. A. Donmez, D.S. Blomquist, et al, "A General Methodology for Machine Tool Accuracy Enhancement by Error Compensation," Journal of Precision Engineering (1986).

11. K.W. Yee and D. Blomquist, "Checking Tool Wear by Time Domain Analysis," Manufacturing Engineering, 88(5), p. 74-76 (1982).

12. J.A. Simpson, "National Bureau of Standards Automation Research Program" in "Information and Control Problems in Manufacturing Technology, 1982," (D.E. Hardt, ed.), Pergamon Press, New York (1983).

13. R.H. Cross and D. Yen, "Management Considerations for Adopting MAP," CIM Review 7, p. 55-60 (1990).

14. J.S. Albus, "Brains Behavior and Robotics," BYTE Books, McGraw-Hill, New York, NY (1981).

15. E.J. Barkmeyer, J. Lo, "Experience with IMDAS in the Automated Manufacturing Research Facility," NISTIR 89-4132 (1989).

16. J.S. Tu, T.H. Hopp, "Part Geometry Data in the AMRF," NBSIR 87-3551 (1987).

17. L.P. Sullivan, "The Power of the Taguchi Method," Quality Progress, p. 77-82 (1987).

CONCURRENT ENGINEERING
THROUGH PRODUCT DATA STANDARDS

GARY P. CARVER

HOWARD M. BLOOM

Factory Automation Systems Division
Manufacturing Engineering Laboratory
National Institute of Standards and Technology
Gaithersburg, MD 20899

I. INTRODUCTION

Product data standards will revolutionize U.S. manufacturing and enable U.S. industry to build on its traditional strengths and regain its competitive edge for the twenty-first century. Standards will enable concurrent engineering to be utilized in the diverse, dynamic and heterogeneous multi-enterprise environment that traditionally has characterized U.S. industry.

Concurrent engineering provides the power to innovate, design and produce when all possible impacts and outcomes can be considered almost immediately. It is the use, in all phases of a manufacturing activity, of all the available information about that activity. It represents the commonality of knowledge applied to a production goal.

Concurrent engineering can stimulate and maintain the diverse and individualistic nature of the entrepreneurial environment by expanding access to knowledge. It forces a global optimization among all of the product life cycle processes within a design and production system.

However, in an automated environment, concurrent engineering is impossible without standards. That is, the full automation and integration of industrial processes is impossible unless standardized hardware and software, especially standardized knowledge and knowledge models, exist to allow intercommunication among all types of computerized systems. The significance and potential impact of this assertion are the subjects of this chapter.

In principle, concurrent engineering does not have to be an automated process; it could be people interacting directly with other people. In practice, in today's manufacturing environment, the increased complexity of products and processes and the use of computerized systems precludes sole reliance on people-to-people concurrent engineering. The approach to concurrent engineering has to be through the automatic sharing of knowledge by computerized systems. It can be thought of as automated concurrent engineering, or computer-aided concurrent engineering.

In the U.S., the introduction of concurrent engineering to an enterprise, or to a group of connected enterprises, through people-to-people interactions requires usually unacceptable cultural changes. Because it emphasizes teamwork rather than competition, people-to-people concurrent engineering may be in conflict with a company's culture or management style. Or it may interfere with established relationships among the departments within a company or among the companies within a group of companies.

However, introducing concurrent engineering through integrated computer systems does not require cultural changes. Even while the integrated computer systems are sharing information, people in the manufacturing environment have the choice of how they respond to the information presented to them automatically by their computers. They do have to alter the way they work because they are utilizing greater amounts of information; however, they do not have to alter the way they interact personally with other people. In this manner, concurrent engineering does not require cultural changes. People and companies can interact and can perform their activities either individually or collectively, whatever style suits them. The entrepreneurial spirit does not have to be stifled by business-imposed interactions. The key is that the computer systems used by the people and companies interact effectively, and automatically.

Concurrent engineering achieved through the integration of computer systems can create a cooperative environment within a company, as well as among companies. In fact, "multi-enterprise concurrent engineering" can result in bringing together independently innovative companies without any loss of independence. This will provide the mechanism for the U.S. to develop its own, unique, U.S. culture-based approach for achieving world-class manufacturing.

If the approach to concurrent engineering is through automation, concurrent engineering requires the application of information technology to create the means for automated systems to communicate and interoperate. For example, within a manufacturing enterprise, computer-aided design systems must be able to share information with analysis systems, manufacturing systems, and distribution systems. Eventually, concurrent engineering can be applied to all business systems, not only manufacturing systems.

Interconnected automated business systems will provide managers, engineers, accountants, marketing specialists, distributors, and everyone involved in a business enterprise with all the information they need to carry out their functions. This includes information they need to make decisions as well as information about how their decisions affect the decisions and activities of everyone else in the business. Plans and actions will be made simultaneously, without the delays experienced in traditional paper communications and face-to-face meetings as projects progress step-by-step in linear fashion.

Even suppliers, partners, and customers can be linked through an information network. In this way, multi-enterprise concurrent engineering can create vertically or horizontally integrated manufacturing entities *de facto*. Although the suppliers would not be controlled by the systems integrator, for example, as they might be in a vertically integrated entity, supplier companies and systems assembly companies might cooperate to their mutual advantage through the sharing of product data and decision-related information.

In our increasingly global economy, digital information technology has emerged as a critical determinant of international competitiveness. From computers to telecommunications to military systems to consumer electronics, the future of a nation's economic and worldwide influence will depend on its excellence in digital information technology. Just as the industrial revolution changed the world order, the information revolution will too. Just as steel, ships, and computers affected the balance of economic and military power, information technology will too. Concurrent engineering is one of the applications of information technology that will provide unique economic opportunities.

The result of multi-enterprise concurrent engineering is more than just the optimization of a design and production system--it is a broader optimization of an industrial system. The technical challenges are numerous and difficult.

Equally challenging is the attainment of international consensus on the methods for achieving the required networking of diverse types of business systems. International consensus on the means for integrating automated systems--the standards--is essential. No single company, in fact no single country, has enough resources to develop suitable methods applicable to all businesses in all countries. Even if it were to happen that one company developed an integration method, the likelihood of acceptance by everyone else is negligible. Clearly, the best approach is through consensus-based international standards.

Yet sometimes standards are viewed as constrainers of innovation and inhibitors of new technologies. Fortunately, standards for enterprise integration are interface standards or "open system standards." Interface standards relate to interoperability, including data exchange and intercommunication, among different hardware and software elements. Interface standards encourage independent development of interoperable products because they specify both the characteristics of critical interfaces and the way in which the information transferred across the interfaces is represented digitally.

Such standards are welcomed by manufacturers because they lower barriers to market entry and they enlarge the market. From the customer's view, open system standards lead to more intense competition, a larger number of vendors from which to choose, a greater variety of off-the-shelf solutions that are both less likely to become obsolete and more likely to be easily integrated into existing systems, modular systems that can be configured for improved performance in a specific application, and, as a result, lower prices. Manufacturers do not want to venture down proprietary paths with the risk that they may one day find themselves at a dead end. Everybody wins.

Concurrent engineering, based upon product data standards and enterprise integration framework standards, truly represents a new form of concurrent engineering that can be called "multi-enterprise concurrent engineering." Multi-enterprise concurrent engineering extends the principles of concurrent engineering to our U.S. environment. It can be defined as the systematic approach, across industrial enterprises, to the integrated concurrent design of products and their related processes (such as manufacturing and support) through the sharing of product data.

Achieving the benefits of concurrent engineering (Section II) requires an understanding of the unique role of the design process in the life cycle of a product (Section III). However, concurrent engineering encompasses more than the individual processes in the life cycle of a product. It also includes social practices and customs among people and their organizations that are involved in those life cycle processes. Fortunately, it can be shown that concurrent engineering practices implemented using integrated automated systems will not interfere with traditional social interactions but will greatly enhance the strengths of the U.S. style of commerce (Section IV). The essential ingredients for this to happen are the technologies and standards that will allow the sharing of information among all computerized business systems. An unprecedented effort by a variety of organizations to develop the required technologies and to implement internationally the required product data standards is underway. STEP, the Standard for the Exchange of Product Model Data, is the focus of that effort (Section V).

II. CONCURRENT ENGINEERING IS TEAMWORK-IN-EFFECT

Concurrent engineering is a process that involves the integration of information. In principle, in a concurrent engineering approach, all the available information about a product is accessible at every stage in its design, manufacture, support, and recovery or disposal, as illustrated in Figure 1.

A concurrent engineering approach could be implemented by assembling a team of (human) experts, each of whom is a specialist responsible for one or more stages of the product's life cycle. The team would create and support the product over its life. Access to information would be accomplished either by request for

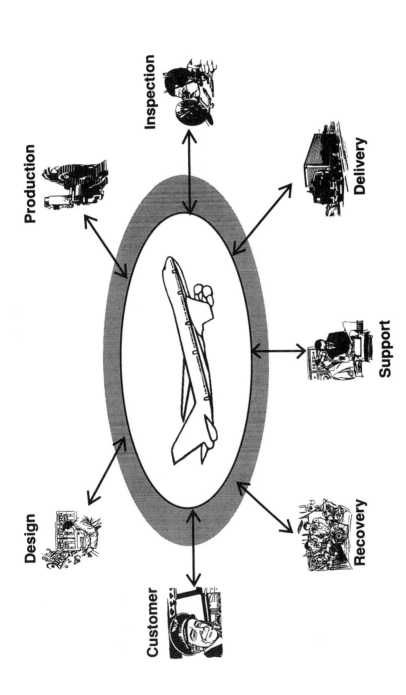

Figure 1. A "world class" product requires an enterprise-wide commitment to excellence. Examples of various life cycle stages of a complex product--an airliner--are shown schematically.

the information by the expert who recognizes the need for it or by contribution of the information by the expert who recognizes its usefulness at the time. The human team is the mechanism for integrating the product information. Examples of product information used by experts in different phases of a product's life cycle are shown in Figure 2.

The human team approach works, but is limited by cultural and organizational practices and by the amount and complexity of the information. An alternative approach that has none of these limitations is automation of the creation-to-disposal process for a product. In this approach, computerized systems access the information and either automatically utilize it or offer it to the appropriate human specialist at the proper time. Therefore, the mechanism for integrating the product information is the totality of interconnected, information-sharing automated systems. This is illustrated in Figure 3. No human team need meet or interact face-to-face. The specialists could be separated both physically and organizationally.

Of course, human specialists must still play a role. They create designs, make value judgements, and make decisions from information provided by the automated systems.

Nevertheless, concurrent engineering truly represents teamwork, even in its automated embodiment, when an actual team of human experts has not been created. Using a concurrent engineering approach made possible by integrated automated systems, product experts operate as they would in a traditional environment. This is because they (or their computers) utilize information from and provide information to each other as needed.

A. The Meaning of Concurrent Engineering

Concurrent engineering is an old concept. It has sometimes been called concurrent design, simultaneous engineering, and system engineering. Even prior to these labels, in earlier times when individual craftspeople created individual objects, they took into account such factors as the properties of the materials, the manufacturability of the parts, and the function and utility of the object. The integration of the product information occurred within the mind of each individual craftsperson. The end result was a complete product, ready for use by the customer.

When factors such as technology led to more complex products as well as specialization and compartmentalization among experts and workers, the integration of all relevant information was no longer spontaneous. Increasingly, the tendency was that the information was made available *sequentially*: the designer designed, then the manufacturer manufactured, and so forth.

In contrast, concurrent engineering is an inherently *parallel* process. The integration of the information required for all phases of the life of a product represses serialism and promotes parallelism. A formal definition of concurrent engineering emphasizes this idea [1]:

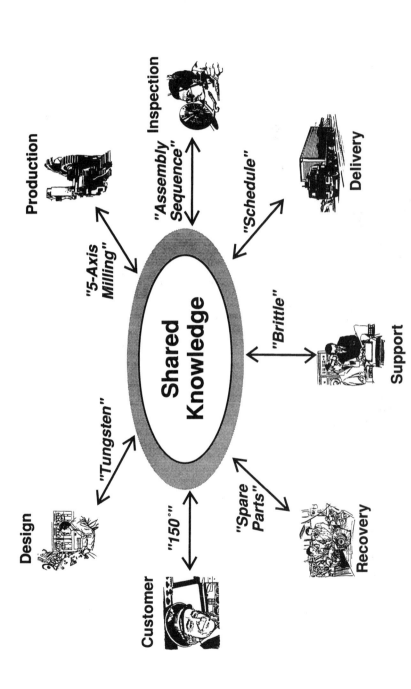

Figure 2. Examples of knowledge that is needed by the specialists involved in the various product life cycle stages are illustrated. The sharing of the product knowledge leads to concurrent engineering.

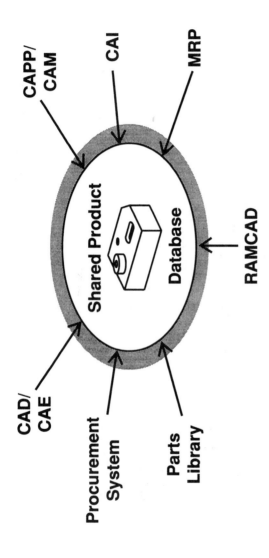

CAD/ CAPP/
CAE CAM CAI MRP

Procurement
System

Shared Product

Parts
Library

Database

RAMCAD

CAD — Computer aided design
CAE — Computer aided engineering
CAPP — Computer aided process planning
CAM — Computer aided manufacturing
CAI — Computer aided inspection
RAMCAD — Reliability and maintenance CAD
MRP — Materials resource planning

Figure 3. When the knowledge is shared automatically by computerized systems, the concurrent engineering environment itself becomes automated.

"Concurrent engineering is a systematic approach to the integrated, concurrent design of products and their related processes, including manufacture and support. This approach is intended to cause the developers, from the outset, to consider all elements of the product life cycle from conception through disposal, including quality, cost, schedule, and user requirements."

It is apparent that in today's U.S. manufacturing environment barriers must be overcome to realize concurrent engineering. Therefore, in practice concurrent engineering may mean:

- Overcoming resistance to teamwork; that is, getting designers, manufacturing engineers, and support personnel to work together,
- Overcoming competitiveness between individuals,
- Retraining the educationally specialized in newer technologies,
- Modifying management styles and organizational cultures, and
- Developing new types of computer-based tools.

These are input- or investment-oriented issues. Looking instead at what concurrent engineering means in terms of outputs or benefits, concurrent engineering may mean:

- Lower costs,
- Shorter time-to-market, and
- Greater quality.

These results will affect the survival of a company or the success of an industry in world markets.

While any and all of these issues are real (and are explored later in Section IV), the single most important issue regarding concurrent engineering is that standards for all types of enterprise information--especially product data standards--are essential because *concurrent engineering is impossible without standards*.

Even if a multidisciplinary team of engineers were assembled to produce a product as well as the procedures necessary to maintain it in use, it is impossible such a team could operate effectively without the ability of the automated systems to communicate and share information about all phases of the product's existence. In today's manufacturing environment, concurrent engineering means integrated information systems--and that means product data standards are needed.

B. The Need for Concurrent Engineering

Competitive success depends on shortening the time between conception and introduction of new technologies and products into the marketplace. To meet the need to minimize time-to-market, computer-aided tools are used to move the product from concept through design, prototype, manufacture, test, and introduction into the marketplace (from concept to consumption).

Even as pressure is applied to decrease product development time, the diversity of activities and expertise required to bring a product to fruition are increasing dramatically. This is because in addition to meeting functional needs, the product must meet energy, environmental, health and safety, and other requirements. These non-traditional requirements are becoming more demanding on manufacturers as the sophistication of our culture advances and as our knowledge of the impacts of human activities on the global environment expands.

Because of the need to minimize the time from conception of a product to its delivery to a customer, and because of the amount and variety of information needed by manufacturers, computers are essential in manufacturing. They are used to design products, plan for their manufacture, control the equipment that produces them, control the equipment that tests them, manage their distribution, and help support their operation, repair and maintenance. Furthermore, because most manufacturers of complex products, for example, vehicles and computers themselves, manufacture only a fraction of the parts in these products, there are needs relating to activities such as inventory control, scheduling, and ordering, as well as the coordination of all the manufacturing activities of the supplier companies. But the needs can be only partially met, and the advantages of using computers only partially achieved, unless the computers can interoperate--that is, share information--so that they can perform their tasks in parallel. In this manner computers and integration naturally point to--even *demand*--concurrent engineering.

C. The Benefits of Concurrent Engineering

Concurrent engineering can shorten the time required before a product is marketed. It can improve productivity, profitability and competitiveness. It can lower costs, reduce waste, improve quality, and improve efficiency in all phases of the life of a product [1]. It can allow suppliers and vendors to coordinate their operations. It can enable manufacturers to cooperate in consortia in precompetitive projects. Such coordinated operations increase the resources, lower the cost and reduce the risk for each individual participant. This may allow the participants as a group to be more innovative and risk-taking to produce more competitive products. In today's aggressively competitive world market, concurrent engineering may mean the difference between failure and success of an industry.

The economic benefit of success (that is, at least survival) compared to failure is obvious. But there are other benefits related to overcoming the barriers:

Teambuilding: Concurrent engineering integrates, through the sharing of information, all the people involved in a manufacturing activity. They become a team through their automated systems. The interactions can be optimized, even though they may not include face-to-face interactions, and the individual team members can perform to the limits of their capabilities without changing their personal or interpersonal styles.

Workers' Career Growth: New forms of workplace organization in modern manufacturing facilities gives workers more responsibility [2]. Increasingly, they must use judgement and make decisions. Concurrent engineering will accelerate this trend, and it will support workers with the information they need.

Management: The concurrent engineering environment will give managers the ability to oversee all activities throughout their enterprise--without any additional bureaucracy. Managers will be able to have the information they need, as it is created, to anticipate, plan, and act quickly.

Competition and Choice: It takes years for companies in an industry to recognize all of the different specialty niches for systems and to develop viable products. For example, although the basic interface specifications for personal computers were established in the early 1980's, new types of hardware and software products are still being defined today. In the same way, the integrated concurrent engineering environment will provide opportunities for new products that cannot be predicted now. The enabling vehicles are standards. In addition, since no one manufacturer offers computers that will design, manufacture, and support a product, interface standards are essential.

Additional thoughts on the benefits of concurrent engineering as both a method and an environment are discussed in Section IV. In the next section, an important reason why concurrent engineering can provide considerable benefits is discussed. The reason is that concurrent engineering affects design decisions. Concurrent engineering, based upon integrated automated systems, allows specialized engineers from all phases of a product's life cycle to participate in design decisions, and this has a major impact on the cost of the product [3].

III. DESIGN IS THE CRITICAL ARENA FOR CONCURRENT ENGINEERING

If manufacturing is the use of energy to convert materials and components into saleable products, then design is the use of knowledge to convert information and requirements into functionality. Design includes both design of a product as well as design of the manufacturing systems and processes to produce the product. Design decisions affect all aspects of the life of a product, including production cost and other characteristics of the product's manufacture, marketing, maintenance, repair, and disposal. A "good" product design addresses the concerns of each of these characteristics as well as the quality, cost, and functionality of the product to the user. Examples of the kinds of information needed by a designer are listed in Figure 4. World-class products, products that are competitive in timeliness, performance, cost, and quality, result from good design.

The term "concurrent design" refers to the "coordinated design of products and processes so that effective and efficient manufacturing will be possible" [4]. Concurrent design is therefore consistent with intuitive concepts of "good" design. (Although good engineering design has been more formally described as a process that ensures "products are designed and manufactured with 'designed in' instead of 'tested in' reliability and maintainability" [5].) The point is that concurrent design is the only design approach that works in a concurrent engineering approach.

A. The Uniqueness of Design in the Life Cycle of a Product

Possibly because it includes the initial stages of the development of a product, and certainly because it determines the nature of the future attributes of a product, the process of design exerts the most control over a product's life cycle. For example, about 60% of a product's cost is fixed very early in the process of design; overall, the design process may fix as much as almost 90% of the total cost of a product [6]. This means that production and production management decisions affect only about 20% to 30% of the total costs. This is shown in Figure 5. Despite its importance, the design process is often inefficient, detached from the production process, undocumented in terms of the rationale for design decisions, and production-facility dependent.

A fundamental problem with developing effective design environments and representing design intent has been a lack of a conceptual model of the design process. Without a model for the various facets of the design process, computer-based design tools will remain customized and relatively isolated, interacting only at low levels. In this situation, unique solutions and individualized integration schemes substitute for an appropriate architecture for the use of concurrent engineering in design.

Design

The product designer must develop a design that meets the basic conceptual requirements for the product while satisfying manufacturing, operational and maintenance constraints.

Operational Constraints

· Product use/mission
· Functional flexibility
  · Range
  · Payload
  · Operating conditions
· Service cycle time
· Cost per unit

Manufacturing Constraints

· Precision/tolerances
· Process equipment and tooling
· Special handling
· Hazardous materials
· Clean room conditions
· Staff skills required
· Production Time
· Cost of materials

Maintenance Constraints

· Technician skills and training
· Ease of access
· Special tooling or equipment
· Ease of problem diagnosis
· Modular assemblies
· Special handling
· Hazardous materials

Figure 4. The designer especially has a need for knowledge about the properties of a product during its life cycle.

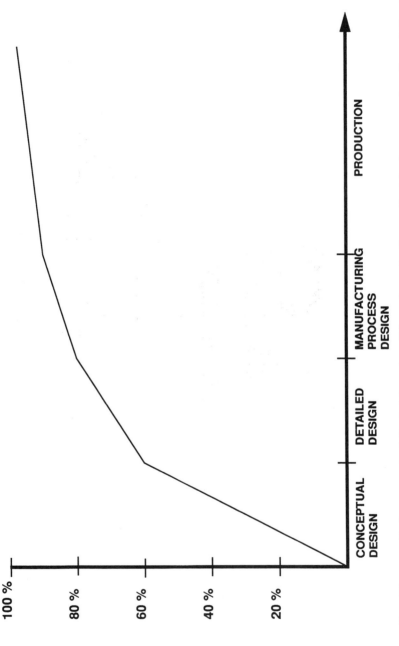

Figure 5. The cumulative percent of the total cost, or life cycle cost, of a product is shown as a function of the stages of the product's life. Early design decisions determine most of the ultimate cost. (Adapted from [6].)

The design process can be divided into conceptual design, detailed design, and manufacturing system/process design. Early in the conceptual design phase, just when there is the most flexibility, most of the total costs of the product are committed.

1. Conceptual Design

The conceptual design phase is when the concept for the product, including sub-assemblies and components, is developed. The general shape of the product is known, but detailed geometric information is not yet available. The output from conceptual design activities contains assumptions, constraints, conditions, and other information related to a product that must be used by downstream operations to produce and support the product.

Important decisions are made during conceptual design that affect the nature of a product, such as complexity and maintainability. The computer-aided engineering (CAE) tools required to involve design in the overall concurrent engineering approach should be integrated and interfaced and should be capable of representing functional information about the product. The designer needs such information early in the design process, and in a form that is readily accessible to the design tools being used.

Computer-aided engineering tools allow the designer to model the qualitative and functional performance of the product. CAE tools include tools for simulation of operations, structural and mechanical analysis such as finite element and boundary element analysis, fluid flow and thermal analysis, solid and surface modeling and a variety of other specialized evaluations of the potential product. Such analyses to determine performance characteristics are executed usually on an idealized geometry. Unfortunately, most of today's CAE tools have their own proprietary data structures and interfaces. The data are not available to other computer programs and are interfaced only to the human user.

In addition to lacking the ability to exchange information, most of today's CAE tools also lack the ability to represent non-geometric functional information about the product. It is important to know, for example, whether a product must be non-conductive or corrosion resistant and whether it will be used in conjunction with or interact with some other device. To allow for automated or computer-sensible concurrent engineering, CAE tools must be able to capture function information and to represent physical objects better than is possible today.

2. Detailed Design

During the detailed design phase the product is specified completely and unambiguously. The geometry and topology, the dimensions and tolerances, and any features needed as a result of the analyses performed during the conceptual design phase are specified in detail.

In addition to specifying the product in detail, assurance must be developed that all features of the product design are consistent with priorities and considerations for downstream life cycle stages. For example, it must be determined that tolerances are neither too loose to meet functionality needs nor too restrictive for manufacturability and low cost. Tolerance decisions reflect the capability of available manufacturing equipment; either the designer modifies the design to accommodate equipment-dictated tolerances or designs-in tighter tolerances that necessitate specialized equipment. To create a design having predictable product functionality, manufacturability and cost, the designer must know the effects of alternative design decisions.

The designer also needs information about the importance of the product attributes as they relate to the processes the product will encounter in its life to make knowledgeable design decisions. Questions such as, "Is performance more important than maintainability?," "Is ease of assembly more important than durability?," and "How do these characteristics relate to cost?" must be considered and must have answers if detailed design is to be completed successfully.

In a concurrent engineering environment, the design tools have access to information about downstream life cycle considerations, including costs. Furthermore, the tools are able to measure and express in a quantifiable manner the results of analyses of the impact of detailed design decisions on downstream stages of the life cycle. Obviously, the CAE tools must be interfaced to other automated systems if they are to accomplish such tasks.

3. Manufacturing System and Manufacturing Process Design

The design of the manufacturing systems and processes that will be used to produce the product involves detailed knowledge of a different type than may have been needed earlier in the design process. Information such as the number of devices to be produced, the batch size, the process capabilities, the availability of materials, and other information about manufacturing operations and equipment was needed earlier, but typically only to avoid a design that violated certain practical limitations on its production. At this point, this kind of information is needed to determine the actual sequence of manufacture and the facilities and production equipment required.

The CAE tools used at this point are also different than those used earlier in the design process, but they share the same need for integration and interfacing. Software for process planning, modeling of the actual manufacturing processes, programming and controlling the production machines, fixturing the machines, and other production operations must have access to product design information.

It is easy to see how tolerance information would be important because it relates to the production processes and their capabilities for precision. What is less easy to understand is how design intent information is critical for concurrent engineering. This is considered in the next section.

B. The Representation of Design Intent

Solid modeling systems improved the design process when they replaced "wireframe" drawings. Solid modeling systems can be used to represent an object being designed accurately and unambiguously. Recently, solid modeling systems have been used to generate finite element meshes and to generate tool paths for machining. However, solid models still do not help in all stages of the product life cycle. They only help to provide more complete geometric information. The goal for concurrent engineering is product models that include not only dimensions and tolerances, as well as other feature information, but also information about the decision-making process that led to a product design.

To maximize the benefits of concurrent engineering, the various product life cycle systems must be able to determine how, for example, the shape or dimensions were derived and why the part must be non-conductive. Unless such design-intent information is available, it is difficult to minimize the risk of making modifications and improvements to products and manufacturing and repair processes.

New software tools and models are needed to measure such properties as design "goodness;" manufacturability; product and process performance; product and process costs; impacts of changes from one alternative design to another; and manufacturing configuration. The following are some examples of needed design tools and models [1], [7]:

Process models for various manufacturing processes such as metal cutting, forming, injection molding, and casting: The models can be used to alter product geometry or process parameters through various engineering, geometric, statistical, and scientific analyses. The models can provide a measurement (such as tolerances, material integrity, strength, and surface finish) for projecting process performance (such as speeds, temperatures, and precision).

Assembly and cost models to provide the designer with cost prediction estimates and design options for more easily assembled products: Geometric models of position and path for product handling are also useful.

Manufacturing system models to measure the capacity of systems: The integration of capacity and capability models in manufacturing allows the designer to simulate and analyze the manufacturing processes, the process technologies, and the design of statistical quality control methods.

Factory engineering models to provide the data required to configure new factories: These tools can be used to show how a decision to improve one process in the factory might result in simultaneous changes in the data maintained in several associated applications. Some changes which might occur automatically include revisions to plant layout drawings, utility requirements, simulation models, cost/payback projections, and procurement specification documents for the proposed system.

Future systems will be "intelligent machines" that receive product descriptions and automatically interpret them and perform the appropriate machine operations to achieve the desired product geometry, tolerances, and material specifications. The keys are the integration of the design database into the manufacturing facility and the integration of methods for translating design descriptions into manufacturing process and control programs.

The process of design, as well as manufacturing generally, will be improved by the development of standard ways for representing design intent. Software tools will need to include a better understanding of the design process. The goal must be the inclusion of product data models that incorporate design intent information. Only with a standard representation for such information can the output of design systems be used by the variety of other automated systems responsible for a product during its life.

It should be clear that design is a process that involves a series of activities. During the design process, a large problem is decomposed into a series of smaller problems, partial solutions are proposed, and, through feedback and the resolution of constraints or tradeoffs, partial solutions are refined until an overall solution is reached.

Present computerized aids merely help speed up the design process by performing quickly the computations needed for the partial solutions. To achieve a concurrent engineering environment, a framework based upon a model of the design process and appropriate interface and product data standards is required for the integration of computer-based tools. One approach is to view the concurrent engineering environment as a "large-state machine" within which changes in the state of the system can occur in predetermined ways [8].

IV. CULTURE AND CONCURRENT ENGINEERING CAN BE COMPATIBLE

Our current information-based society [9] seems to be characterized by the computerization of everything. Early in this period, some people feared that computerization was equal to dehumanization. However, it is now clear that it is the way computers and digital systems are implemented that can either limit or extend human interaction and control or that can either diminish our rights or endow us with additional freedoms.

Computers seem to be the most visible sign that as the world is changing, cultural and traditional values are coming into conflict with technology. Yet there is evidence that computers can be used to preserve or even restore traditional approaches--at least at the human-machine interface. In a fundamental sense, the term "user friendly" implies a recognition of and deference to human ways of doing things. The concept of "artificial intelligence" connotes an imitation of the human mind's ability to process information. These are examples of computers helping to extend our behavior and abilities.

The relevant question is: "Does concurrent engineering represent an attack on human cultural values--such as creativity, individualism, and pride--or does concurrent engineering allow us to extend our abilities and do more of what we did before, in the way we do it best?"

The answer is that concurrent engineering is in harmony with us as humans and with our culture. Furthermore, concurrent engineering may be the only way to preserve our style of doing business in the U.S.

A. The Relationship Between Concurrent Engineering and Competitiveness

The U.S. loss of competitiveness, shrinking markets, and trade deficits are symptoms that something new is going on in the rest of the world. There can be little doubt that today the U.S. is losing world market share and technological leadership in most of the technologies important to maintaining our standard of living [10]. According to a recent government report on electronics, a major growth area in the U.S. economy in terms of employment, output, exports, and innovation and also vital to national defense and security, U.S. leadership is under serious challenge and may soon be eclipsed [11].

Another government report begins a summary in chapter one with the declaration: "American manufacturing has never been in more trouble than it is now" [12]. According to this report, manufacturing is weak because its technology is weak. And unless this weakness is cured, the U.S. will not be able to enjoy rising living standards and the continued creation of jobs at the same rate as in the past. This includes not only jobs created in the manufacturing sector, but also indirectly in the service sector. Manufacturing technology is identified as the key to national competitive success.

There can be little doubt that in today's economic climate, and for the foreseeable future, industries must compete in world markets. Among industrialized nations, international competition is increasing. Yet, as evidenced by the trade deficit, many U.S. industries are not competing successfully.

A number of factors are cited that contribute to the current difficulties of U.S. manufacturers in global markets. The factors often include federal budget deficits, low personal savings rates, the high cost of capital, the low "patience of capital," the weak dollar, short-term profit goals of corporate managers, short-term profit goals of investors, lack of a trained workforce, lack of access to foreign markets, "dumping" by foreign companies, product liability and litigation, inadequate foreign protection of intellectual property rights, federal regulatory restrictions, poor management-labor relations, foreign manufacturer-supplier relationships, antitrust laws,... Seldom, if ever, do the reasons given for U.S. poor competitiveness include either technology or standards issues. This may be because the role of technology and its relationship to economic factors may not be well understood. Technology is sometimes viewed as only the creation of new products rather than the improvement or revolutionary changing of existing manufacturing practices. As mentioned in the introduction, standards are sometimes viewed as inhibiting change rather than enabling or facilitating innovation and competition.

Technology can make a difference when it provides the means to build a new national economic strength or to build on an existing strength. In these ways technology can provide the means to overcome non-technical barriers and to produce a new basis for competitiveness. Completely new technologies can even change the significance of the non-technical barriers, making some of them inconsequential. Concurrent engineering based on integrated automated systems may be able to do both.

Actually, concurrent engineering has already been shown to be able to contribute to competitiveness. A Harvard Business School study showed concurrent engineering to be responsible for a 30% decrease in the time-to-market for a new car in the Japanese automotive industry [12], [13]. Even in the U.S., in a study of six defense contractors, concurrent engineering has already been shown to reduce costs 30% to 60%, to reduce development time 35% to 60%, to reduce defects by 30% to 80%, and to reduce scrap and rework by 58% to 75% [1]. However, concurrent engineering, along with integrated computerized systems and appropriate standards, can make even more significant improvements.

B. The Relationship Between Concurrent Engineering Practices and Alliances Among Businesses

Manufacturers are linked in a chain, sometimes called a "food chain," to their materials and parts suppliers and to their customers. It would seem that close links and stable relationships would be advantageous.

In 1988, a deteriorating trade balance between the U.S. and Japan reached a deficit of $6 billion. In 1989, for the first time, the U.S. was a net importer of computers. In 1990, also for the first time, the computer systems industry had a zero trade balance. One important reason for these trends may be cultural. The perception is that while U.S. industry is considered to have better design skills, it does not have the business partnerships and long-term strategies for commercial success. The Japanese, who lead in rapid and integrated "design for manufacturability" and in flexible manufacturing, have strong and stable business partnerships as well as long-term strategies.

It has been argued that the advantages of Japanese companies in world markets are due primarily to their diversified, vertically integrated structure and their long-term partnerships in financial-industrial groups called "keiretsu" [14]. Typically, a major bank is one member of a keiretsu. This is important because today the size of the investment necessary to develop a new technology is prohibitive to most companies. The ability of a company to develop a technology and quickly bring to market a product that utilizes that technology is an obvious advantage. This can happen when a partner company with strong consumer product development skills and effective marketing networks develops the product even as the technology is being developed. In a sense, this is multi-enterprise concurrent engineering.

Yet given the evidence, U.S. industry has been slow to form cooperative relationships. Although more companies are forming consortia, especially for joint research and development in precompetitive technologies, and more companies are cultivating closer relationships with their suppliers, there is still resistance to business alliances in U.S. industry. This is true even though the interpretation of anti-trust regulations is gradually weakening and federal funds are becoming available for consortia in generic and precompetitive technologies.

The individualistic entrepreneurial spirit helped bring us to where we are today. But unless something changes, that same characteristic will prevent the U.S. from reversing its decline in competitiveness. This is because in a world "where R&D and commercialization costs can exceed $1 billion for a single technology, small companies working independently cannot compete. [The fragmentation of U.S. industry into small companies]...spreads R&D funds thinly, slows technology diffusion, and diffuses manufacturing and marketing power" [15]. Yet U.S. technology companies do not make the strategic partnerships that Japanese companies do and, if the current trade, market share, and direct investment (Japanese companies buying high technology U.S. companies) statistics continue, small innovative U.S. technology companies may not be able to survive.

The behavior of U.S. companies reflects their cultural environment. "Rugged individualism" is a desirable characteristic. Strengths include creativity, innovation, and individualism, even--perhaps especially--in commerce. Competition has been a primary motivator and strength of our system; but it does not coexist well with the notion of entrepreneurs working together. The "not invented here" syndrome and the stereotypical "throw it over the wall" style among departments within companies is symptomatic of cultural values. In addition, the tendency toward an adversarial relationship can divide management and labor.

How can all this be reconciled with the need to cooperate to form large, internationally competitive business structures? How can it possibly support concurrent engineering?

The answer is that the creation of an automated interconnected computer environment can free people and companies to operate however they please. If information can be shared automatically and the computer systems people use to do their work can interact and the mechanisms are in place to create a concurrent engineering environment, then individuals can perform their activities in their own way. Instead of being constrained by having to interact and cooperate with others, people can, if they choose, be independent and work alone. Their data will be integrated. In this way, the full automation of the industrial environment will provide the needed teamwork. The entrepreneurial spirit does not have to be stifled by business-imposed interactions. Independent innovators can continue to be independent--at least in the way they operate personally.

Consequently, concurrent engineering and all it represents is not necessarily antagonistic to our culture. In fact, the creation of a concurrent engineering environment through information technology can free people and businesses to do what they do best. In the U.S., concurrent engineering can provide the cooperation that allows us to continue to be independent entrepreneurs. Concurrent engineering can meld a large number of small, highly specialized, vibrant and dynamic technology companies into a position of global competitiveness. It can provide the benefits to industry of vertical and horizontal integration without undesirable restrictions. It may provide a uniquely U.S.-style cooperative business structure.

V. PRODUCT DATA STANDARDS ARE THE KEY TO CONCURRENT ENGINEERING

The critical ingredient for the use of concurrent engineering in manufacturing is the integration of product and process data. This integration gives the designer, along with everyone else in the commercialization chain, information about the entire life cycle of a product, as well as information about how their decisions affect all other aspects of the product. In addition, the integration and automation of product and process data provides for meeting the needs of the specialists involved with a product's commercialization by allowing each of them to obtain a particular "view" of the product that is suitable for their specialty and function.

Within a single enterprise, the issue may be more one of technology than standards. Sometimes a company can choose to use systems from one vendor for all applications, or it can develop the interfaces, translators and other software needed to integrate its systems. But whenever more than one company cooperates or shares information, interoperability of systems becomes the uppermost issue, and that means standards. (Standards do not guarantee interoperability, but they bring it closer to reality.)

The integration of product and process data is not possible unless there is a mechanism that allows the sharing of information among different manufacturing systems. The mechanism must be a standard digital representation for product and process data. That is why product data standards are the key to multi-enterprise concurrent engineering and why, in today's business environment, *concurrent engineering is impossible without standards.*

A. The Path From Automation to Concurrent Engineering

Automation in manufacturing has led to impressive economic benefits from improvements in capacity, productivity and product quality. Yet, the benefits that remain unrealized are even greater. They are the benefits that will accrue from the integration of information among automated systems. The benefits include:

Reduced time from concept to commercialization: The efficient sharing of product data among automated systems will eliminate the need to produce hard copy drawings and models. Design details can be tested electronically against physical and engineering constraints using analysis systems and against economic constraints using cost prediction and manufacturing process simulation systems. Design changes can be made rapidly even after initial production has begun.

Reduced costs: Increased productivity will be obtained by the increased efficiency of the design process (discussed in Section III) and by reduced "time to market." Studies have shown that concurrent engineering results in a reduction in the number of design changes and in the amount of material wasted due to defects and rework [1].

Increased responsiveness to customer needs: By improving the flexibility of the bond between design and production, manufacturers can more quickly introduce new products or change existing products. This capability is essential in today's global markets. The demands by customers for both products and services that are characterized by differentiation, customization, and localization, result in a competitive environment where the rewards go to the speediest and quickest to adapt.

Increased cooperation among suppliers and vendors: The ability to communicate and exchange information among suppliers automatically spreads the benefits of integration from within a single enterprise to a network of enterprises. For example, a single design change in one component can cause an unpredictable delay as its effects cascade through all enterprises whose components and processes are required by the product. Today, even the need to evaluate the impact of a design change on the product causes delays as different enterprises communicate and respond. However, integration would not only allow manufacturers to coordinate activities and product changes among all their suppliers to avoid delays, but, even more importantly, it also would allow manufacturers and suppliers to take any required actions automatically and *simultaneously.*

The integration of information means the merging of machines and information into a system that is responsive and efficient--a system that can support concurrent engineering.

1. A Higher Level of Automation

Typically, many different computer-aided tools require access to computerized product data. The product data represent all the information about the product, including the product's function, its design, the reasons for its design features, and the manufacturing processes that are used to make it. Ideally, the data also describes how the product is to be used or operated, how it is to be maintained and repaired, and how it is to be properly deactivated or disposed of.

The computer-aided tools and computerized information systems that use product data are essentially very large computer programs. They were developed over many years by many people. Because they were developed independently, they tend to use unique representations for storing data. Unfortunately, each system is only able to use data that has been stored in the particular representation that it accepts.

The problem of integrating these systems is the same problem of communicating among people who speak different languages. Each time product information is transferred from one system to the next it must be translated or reformatted. Obviously, having to perform this extra step to share information is inefficient and costly. The costs multiply when many systems are involved. The number of different translators required for a number of different systems to communicate is the square of the number of systems.

The primary reason usually cited for using information systems technology is to reduce costs. Ironically, the incompatibility among existing information systems has the opposite effect: it increases costs. A solution must be found that enables the sharing of product data among different manufacturing information systems. The key to the solution is a standard for product data representation and exchange.

Achieving the goal of concurrent engineering and the economic benefits it represents, is hindered not only by existing incompatibilities but also by the complex nature of the data that must be shared among systems. The manufacturing data that must be shared is more than just an accumulation of unrelated bits of numerical information. Design data provides a good example; it contains physical and functional information as well as information about the significance of the design decisions that led to the final design. Product data includes not only the design data itself, but also data about its supporting infrastructure and its interfaces with other equipment.

Therefore, the standardization of product data implies more than merely the standardization of product data file formats. For concurrent engineering, information models for product data are needed. The path from automation to concurrent engineering is through standardized product data models.

Key to the goal of concurrent engineering is the ability of manufacturing information systems to capture automatically the knowledge that is generated during the product life cycle. The knowledge can then be used by the designer and others involved in managing the product. For example, knowledge about how a product would be processed or what new materials would be required to meet the functional specifications is made available to the designer as the design is being developed to ensure the best quality product reaches the downstream life cycle managers. In this way the most appropriate and cost-effective materials and processes can be used.

In a sense, two attributes of integration can elevate automation to a higher level:

1. The ability to capture automatically the knowledge gained during designing, producing, commercializing, and managing a product throughout its lifetime, and

2. The ability to exchange automatically that product knowledge among different computer systems.

An integrated level of automation can be thought of as a facilitating or an enabling technology for concurrent engineering.

2. Technical Issues in Shared Databases

Automated systems store product information digitally in a database. The mechanism for sharing the information is a multi-user or "shared" database. In a shared database environment, the product information can be accessed by one or more applications, even at the same time.

Of course, there are many technical issues that must be resolved for such a shared database environment to be implemented. As was already described in Section III, an interface is needed between present computer-aided design representations of a product and other computer-aided systems, such as computer-aided engineering analysis and process planning systems. Production costs and capabilities must be integrated into the database for effective design decisions.

Intelligent processes must have access to geometry data in much the same manner that users query business systems. In addition, it is critical to have a mechanism that allows new knowledge to be added to the database as the intelligent processing operations are being performed.

The technical challenge is the development of the information technology and the associated standards that will define the environment for the representation of product knowledge. This will allow the implementation of a shared-database environment for concurrent engineering.

B. The PDES/STEP Effort

PDES, which stands for "Product Data Exchange using STEP," refers to the U.S. activities in support of the development of an international standard for product data sharing informally called STEP, the "Standard for the Exchange of Product Model Data." PDES will help establish a standard digital representation for product data. The specifications already developed by the PDES effort have been submitted to the International Organization for Standardization (ISO) as a basis for the evolving international standard STEP. As the PDES and STEP efforts share common goals, they are sometimes referred to jointly as "PDES/STEP," or simply just as "STEP."

It is important to recognize that STEP is more than a standard for the representation of product data. *The development of STEP is a pioneering effort that includes the research and development of the information technology necessary for the envisioned shared-database environment.* Once this standard and its environment are in place, all types of enterprise information can be more easily shared.

1. A New Approach To Standards

To achieve the goals of the PDES/STEP effort, a new approach to the standards-making process is required. This approach facilitates cooperative development of the requirements, the information technology, and the specifications simultaneously--before the existence of commercial systems that can use the capabilities of the standard.

The technology does not yet exist to define a product and its associated properties and characteristics completely. Even if this could be done now, the technology does not exist to communicate this information electronically and to interpret it directly by the wide variety of automated systems associated with the product's life cycle. Therefore, only a process for creating standards at the same time the technology is being created will succeed for STEP.

The creation of specifications for the standard representation of product data involves many complex issues. It requires a number of different information and manufacturing system technologies and the experience of many different kinds of technical experts. Institutional support for voluntary national and international standards organizations is provided by businesses and industrial consortia and government agencies.

There is a great need for consensus. Industry users and software vendors must cooperate closely throughout the precompetitive technology development and the standardization processes.

In addition, it is essential that the standardization process includes rigorous testing to determine that the standards meet the needs of the user communities. Testing is discussed in Section C, The Technical Challenge of STEP.

2. U.S. Government Needs and PDES/STEP

In 1988, an ad hoc U.S. government interagency task group was formed to focus on information sharing among interested federal agencies. The objectives of the group were to: "prepare and consolidate government requirements for input into PDES development activities, and provide recommendations as to technical and other actions such as needed policy changes, regulatory changes or contractual vehicles/tools (e.g. data item descriptions, contract clauses, etc.) which the government should put in place to foster the development of the PDES specification" [16]. Some of the concerns about the current product data environment expressed in the task group report are:

- It is hard-copy oriented.
- It is massively heterogeneous in terms of vendors and system age.
- Product knowledge is not well captured.
- Product cycles (from R&D to production) are very long and the handoff from one phase to the next phase often loses information.
- Technical data packages are often in error and incomplete.
- Incorporation of changes and technology upgrades is slow.
- New efforts often just automate existing methods.
- Transfer of information to and from contractors is slow.
- Funding for "non-product" development such as PDES is limited and sometimes non-existent.
- Acquisition of improved technology (e.g., new computers and CAD/CAM/CAE) is difficult, time consuming (average 3 to 5 years), and done in the face of ever-shortening technology half lives.
- Industry concern with proprietary data rights is at odds with government desires.
- There is a reluctance for legal reasons to provide CAD/CAM data rather than part drawings.
- Data is replicated many places for different purposes (e.g., non-common/non-integrated databases).

PDES is a major component of the U.S. Department of Defense (DOD) Computer-aided Acquisition and Logistic Support (CALS) program [17]. CALS is tackling a related, but larger scale set of issues:

- Developing and testing standards for digital technical information;
- Sponsoring the development and demonstration of new technology for the integration of technical data and processes;
- Implementing CALS standards in weapon system contracts and encouraging industry modernization and integration;
- Implementing CALS in Department of Defense information system modernization programs.

The emphasis of CALS is the sharing of information by industry and government. The philosophy is that this can only be accomplished through integrated databases that can be accessed by a variety of heterogeneous computer systems, as illustrated in Figure 6. According to CALS, STEP represents the methodology to help accomplish the goal.

Late in 1990, the Department of Commerce and the Department of Defense signed a Memorandum of Understanding (MOU) to "accelerate the development and deployment of technology that will result in higher quality, shorter time to

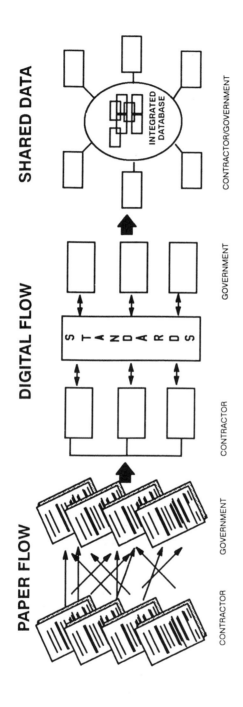

Figure 6. The CALS program is the automation of data processes to transition from paper transfer to digital-information transfer to shared digital information.

production, and lower costs for both weapons systems and commercial products." Cited as essential to this goal are the development of new technology and standards such as STEP. The MOU outlines a partnership program of development, testing, and implementation of standards for product data exchange. The Department of Energy and NASA are expected to enter into similar MOUs with the Department of Commerce in this arena.

3. Institutional Aspects of STEP Development

There are a number of organizations working at both the national and international levels to develop an exchange specification for product data. They include the following organizations:

- IGES/PDES Organization
- ISO TC184/SC4
- ANSI US Technical Advisory Group
- PDES, Inc.
- NIST National PDES Testbed

(The role of the National Institute of Standards and Technology [NIST] and the NIST National PDES Testbed are discussed in the following sections; the other organizations are discussed in this section.)

IGES/PDES Organization. The concept of PDES grew out of the Initial Graphics Exchange Specification (IGES) effort. At the time, the acronym PDES was Product Data Exchange *Specification.*

IGES was first published in 1980 and was updated in 1983, 1986, 1988, and 1990 [18]. Its goal is to allow CAD data to be exchanged between systems built by different manufacturers. When IGES data is passed between design systems, considerable human interpretation and manipulation of data may be required. Since IGES was designed primarily as a mechanism for file exchanges between CAD systems, it is not able to support shared databases between dissimilar product life cycle applications.

IGES developers recognized that a more sophisticated standard would be required to support the integration of different types of product life cycle applications. Therefore, the PDES/STEP effort focused on developing a complete model of product information that is sufficiently rich to support advanced applications, and to support concurrent engineering.

The U.S. voluntary organization that is conducting technical activities in support of the development of PDES/STEP is the IGES/PDES Organization (IPO) [19]. The IPO is chaired by the National Institute of Standards and Technology (NIST) and administered by the National Computer Graphics Association. In 1985 a formal study, called the "PDES Initiation Effort," was conducted. It established a framework and the methodologies for subsequent PDES/STEP activities. Approximately 200 technical representatives from the

United States and other countries meet four times each year to address PDES/STEP-related technical issues.

ISO TC184/SC4. In 1983 a unanimous agreement was reached within the International Organization for Standardization (ISO) on the need to create a single international standard which enables the capture of information to represent a computerized product model in a neutral form without loss of completeness and integrity, throughout the life cycle of a product [20]. In December of the same year, ISO initiated Technical Committee 184 (TC184) on Industrial Automation Systems. Subcommittee 4 (SC4) was formed at that time to work in the area of representation and exchange of digital product data.

Currently, twenty-five countries are involved in the work of SC4. Sixteen of these countries are participating members and nine are recognized as observers. The U.S. is a participating member. The SC4 Chair and the Secretariat are currently held by NIST.

Technical support for SC4 comes predominantly from its working groups (WGs). Alternate quarterly meetings of TC184/SC4/WG level are held concurrently with the IGES/PDES Organization quarterly meetings. Many of the same technical participants from the U.S. and other countries are active in both organizations.

In December 1988, the draft PDES Specification, developed through the voluntary activities of the IGES/PDES Organization, was submitted to SC4 as a draft proposal for the international standard STEP.

ANSI US Technical Advisory Group. The American National Standards Institute (ANSI) is the recognized U.S. representative to ISO and provides the basis for U.S. participation in the international standards activities relating to PDES [21]. To ensure that the positions on standards that are presented to ISO are representative of U.S. interests, a mechanism has been established for the development and coordination of such positions. ANSI depends on the body which develops national standards in a particular technology area to determine the U.S. position in related international standardization activities. Such bodies are designated by ANSI as "US Technical Advisory Groups" for specific ISO activities.

As a participating member in ISO TC184/SC4, the ANSI US Technical Advisory Group (US TAG) selects the U.S. delegates to SC4 and advises the delegates on how they should vote on issues presented to SC4. The US TAG usually meets at each IPO quarterly meeting.

The current US TAG to TC184/SC4 was formed in 1984. Its membership is comprised primarily of technical experts from the IGES/PDES Organization. This type of representation ensures that the technical changes that U.S. engineers and computer scientists believe are necessary are reported to ISO for consideration. ANSI has selected NIST to be the secretariat.

PDES, Inc. In April 1988, several major U.S. technology companies incorporated as PDES, Inc. with the specific goal of accelerating the development and implementation of PDES. The South Carolina Research Authority (SCRA) was awarded the host contract to provide management support. The technical participants provided by the PDES, Inc. member companies and SCRA's subcontractors are under the direction of the PDES, Inc. General Manager from SCRA. At present, there are twenty-four companies that are members, including two foreign companies. Member company's combined annual sales total over $400 billion; they employ over three million people.

PDES, Inc. has embarked on a multi-phased plan for the acceleration of STEP development. Initially the emphasis was on testing and evaluating a data exchange implementation of mechanical parts and rigid assemblies. Current efforts focus on the identification of software implementation requirements, construction of prototypes, and development of "context-driven integrated models" for small mechanical parts. Recently, PDES, Inc. restructured and broadened the program scope to include such areas as electronics, sheet metal and structures. PDES, Inc. is providing increased leadership in the effort to accelerate the implementation of STEP [22]. NIST is a government associate and provides a testbed facility and technical team members to support the PDES, Inc. effort.

4. Nature of STEP

The many different organizations and individuals that are involved in the development of STEP share a common interest:

> *The establishment of a complete, unambiguous, computer definition of the physical and functional characteristics of a product throughout it's life cycle.*

As a standard method for digital product definition, STEP will support communications among heterogeneous computer environments. STEP will make it easier to integrate systems that perform various product life cycle functions, such as design, manufacturing and logistics support. Automatic paperless updates of product documentation will also be possible. The principal technique for integrating these systems and exchanging data will be the shared database.

In the context of STEP, a product may range from a simple mechanical part, such as a bolt or a screw, to a complex set of systems, such as an aircraft, a ship, or an automobile. Ultimately, STEP should be able to represent the information which is needed to describe all types of products, including mechanical, electrical, structural, etc.

STEP addresses many questions about a product: What does it look like? (geometric features); How is it constructed? (materials and assembly); For what function is it intended? (structural and functional properties); How can we tell a good product from a bad one? (tolerances and quality constraints); What are its components? (bill of materials).

The STEP specification is being produced as a series of documents called "parts" [20]. Currently identified parts of the specification[1] are:

- *Introductory:*
  Part 1. Overview

- *Description Methods:*
  Part 11. The EXPRESS Language

- *Implementation Forms:*
  Part 21. Clear Text Encoding of the Exchange Structure

- *Conformance Testing Methodology and Framework:*
  Part 31. General Concepts
  Part 32. Requirements on the Testing Laboratory

- *Integrated Resources:*
  Part 41. Fundamentals of Product Description and Support
  Part 42. Geometric and Topological Representation
  Part 43. Representation Specialization
  Part 44. Product Structure Configuration
  Part 45. Materials
  Part 46. Presentation
  Part 47. Shape Tolerances
  Part 48. Form Features
  Part 49. Product Life Cycle Support

- *Application Resources:*
  Part 101. Draughting
  Part 102. Ship Structures
  Part 104. Finite Element Analysis
  Part 105. Kinematics

- *Application Protocols:*
  Part 201. Explicit Draughting
  Part 202. Associative Draughting
  Part 203. Configuration Controlled Design
  Part 204. Mechanical Design Using Boundary Representation
  Part 205. Mechanical Design Using Surface Representation

---

[1]This list of STEP part titles is current as of February 1991.

The number and titles of the parts are likely to change often as new needs are identified; existing parts may be revised or additional parts may be added to the standard.

There are likely to be many additional application protocols. This is because application protocols are central both to progress in improving and completing STEP and in commercializing it. Application protocols are discussed in the next section.

STEP is defined and represented officially in the EXPRESS programming language [23]. EXPRESS was designed to represent information models in a form processible by computers. The STEP EXPRESS model, called a "conceptual schema," defines and identifies the "objects," or "entities," that can be used by STEP applications. In the first implementation form, STEP product models will be exchanged using a STEP "physical file" [24]. EXPRESS is still being modified and improved to meet the needs of STEP. It will ultimately become an ISO standard (Part 11 of STEP).

IDEF1X is a modeling language that has been used to represent STEP graphically [25]. EXPRESS-G is a newer modeling language extension to EXPRESS. Examples of EXPRESS-G and EXPRESS representations of two simple STEP entities are shown in Figure 7. There are a variety of software tools available for processing EXPRESS [26].

5. Commercialization of STEP

Representatives from industry have key roles in each of the organizations working to develop STEP. Industry must define the requirements for STEP and must assume the most critical role of implementing commercially viable STEP-based systems. After these systems are implemented, industry and government will have to coordinate their efforts to transition jointly from existing information systems to those based upon STEP.

The ultimate objective of PDES/STEP activities is the commercial availability of STEP-based systems. Commercial system developers need:

- Technical specifications that are sound and easy to implement,
- Commercially fair standards that do not favor competitors, and
- A large potential market for their products.

To ensure that STEP is a success, it is necessary that the foundation be built while the specifications are still under development. The problems and issues that will eventually be faced by system developers and users must be identified and addressed before the specifications become standards.

Vendors of systems that will use STEP need to feel confident that the standard is complete, consistent and stable before they will invest in development efforts. Vendors must have easy access to the most current versions of STEP. Help should be freely available to assist their understanding of the standard and how

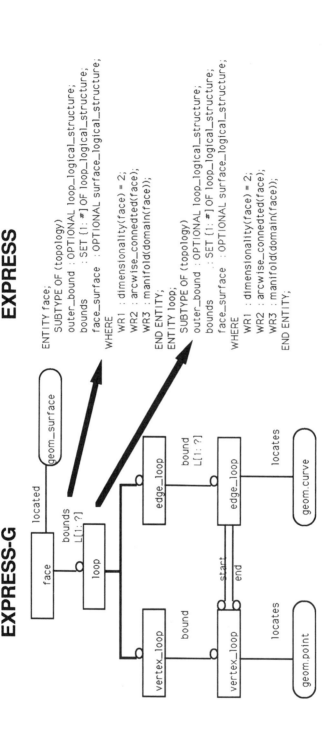

Figure 7. STEP entities in the graphical modeling language EXPRESS-G can be mapped directly to the EXPRESS language format that is used for the standard.

to implement it. Finally, vendors need to know that there is a clearly defined market for systems that employ STEP. Vendors will use application protocols to build their products. For this reason, the strategy is to implement STEP through application protocols, and to extend STEP through the development of new application protocols that bring new entities into the standard as their need is identified.

Application protocols will be standards that define the context, the use, and the kind of product data that must be in STEP for a specific manufacturing purpose in a product's life cycle, such as design, process planning, and NC programming [27]. Application protocols standardize the use of STEP to support a particular manufacturing function reliably and consistently.

An application protocol consists of [28]:

1. An application reference model that specifies the kind of data required to perform a particular purpose, in terms that are appropriate and familiar to experts in the application area,

2. An application interpreted model that defines how the STEP data is to be used to present the information specified in the application reference model,

3. Documentation that describes how the information is used and exchanged, and

4. A set of conformance requirements and test purposes. A corresponding abstract test suite will be developed for each application protocol.

The commercialization of STEP is intimately tied into the development and conformance testing of application protocols.

6. Harmonization Among Different Types of Product Standards

Harmonization, involves the integration, or consolidation, of standards that may be overlapping or conflicting into an unambiguous set of standards that are consistent, compatible, and complementary. Harmonization represents a broad and complex challenge; it must deal with different types of both existing and emerging standards in a variety of industries. Because PDES/STEP development activities are addressing the underlying enabling technologies, PDES/STEP can contribute to harmonization of all product standards.

However, even if there were no overlap or conflict among product data standards, harmonization would still be necessary because complex products include a variety of types of components, and therefore their manufacture requires mechanical, electrical, and other types of data.

Under the auspices of the Industrial Automation Planning Panel of ANSI, an organization called the Digital Representation of Product Data Standards Harmonization Organization was formed to "facilitate the efficacious use of digital representation standards providing a forum for coordination, planning, and guidance to standards developers and approvers" [29]. The harmonization organization has as a long-term objective an integrated set of standards that can support, in digital form, the definition of products for all aspects of their life cycles.

The Organization intends initially to support efforts to integrate four standards sanctioned by ANSI that address the representation and exchange of product definition for electronic products, and to help harmonize them with STEP. These standards, used in electrical, electronic, and electromechanical design and manufacturing, have considerable overlap and conflict and are not consistent with STEP. They include:

- VHDL, "Very High Speed Integrated Circuit (VHSIC) Hardware Description Language," an algebra-like description that is used to design complex logic for chips and computers,
- EDIF, "Electronic Design Interchange Format," a file format for communicating two-dimensional graphics and interconnection information that is often used to describe the patterns that are used to fabricate semiconductor chips,
- Institute for Interconnecting and Packaging Electronic Circuits (IPC) Series 350, used to describe the patterns and mechanical process to manufacture printed circuit boards, and
- IGES, "Initial Graphics Exchange Specification," used to represent the three-dimensional geometry of objects.

Companies that produce electronic products often must use all four of these standards. Unfortunately, since the standards do not work well together, the product information often must be reentered into different computers as the product progresses through its life cycle stages.

The Organization must first define the means for an integrated network of digital product data standards and the definition of a common modeling methodology for all product data standards. (In STEP, EXPRESS is the specification language and IDEF1X is one of several modeling tools.) Once a methodology is accepted, all product data standards models could be integrated into one model. There would also be a common glossary of terms and a dictionary of data entities.

Another aspect may be a structure or "taxonomy" that defines the interrelationships among all product technologies. Such a taxonomy could become a "roadmap" for future standards activities and extensions to existing standards.

C. The Technical Challenge of STEP

There are four major technical challenges facing the developers of STEP:

- The exchange of data is different from the exchange of information. Data must be transmitted accurately and without any changes. In contrast, information, although composed of data, must be understood and interpreted by the receiver. Furthermore, the receiver must be able to apply the information correctly in new situations. The first challenge is that *STEP is a standard for information, not just data.*

- The need for STEP to be extendable to new products, processes, and technologies, requires a more abstract representation of the information than in previous standards. Regardless of their equipment or process, a user must be able to obtain the information necessary to do something from the STEP representation of a product. Therefore, the second challenge is that *the development of STEP must include the development of an "architecture" or a framework for the exchange of information, not just a means or format for storing information.*

- The wide range of industries and the diversity of product information covered in STEP is beyond that of any previous digital standard. The variety of attributes and parameters, such as geometric shape, mechanical function, materials, assembly information, and date of manufacture, is immense. Also, the industrial base, the number of industries involved, is enormous; even greater is the number of technical disciplines that are involved. Moreover, STEP must be flexible and extensible so that new information and additional application protocols can be added and can be upwardly compatible. Therefore, the third challenge is that *the scope and complexity of STEP is far beyond any previous standards effort.*

- Traditionally, standardization is a process that devises an approach encompassing a variety of existing vendors' options, builds on the best solution available, and avoids penalizing some vendors more than others. In the case of STEP, there is no existing implementation. Thus the fourth challenge: *the technology to support STEP must be developed at the same time the standard is evolving.*

The consensus approach to meeting the above challenges is to start with conceptual information models [30]. STEP will consist of a set of clearly and formally defined conceptual models and a physical exchange protocol based on these models. The conceptual models will be combined into a single model with a standard interface to a shared database [31].

The following sections describe the approaches used by the community that is working to develop and implement STEP successfully.

1. Data Sharing

Clearly, it is not the physical hardware connections between computers that is the major issue in data sharing; it is incompatible software. The root of the problem is proprietary data representations, that is, vendor-specific data formats. More often than not, the vendors of computer applications store the data which is required and produced by their systems in their own proprietary format.

For example, once the design of the product has been completed on a CAD system, it is stored in a data file. Some of the information in that data file represents the shape and size of the product. In an integrated information systems environment, the designer should be able to send that data file over to the manufacturing planning system. The same data would then be used by the planning system to determine manufacturing processes for the product, based in part on its specified shape and size.

If the planning system can read the contents of the design data file, it can obtain the shape and size information it needs. It might be said that these two applications are integrated. But, it is a fact today that if two commercial products are integrated, it is likely that they were developed by and purchased from the same vendor. Furthermore, it is also likely that they were intentionally designed to work together from their inception. Often, it is the case today that applications offered by the same company are not integrated.

STEP is intended to address the issue of product data sharing between different computer applications running on different computer systems within the same or different organizations. STEP will provide a standard, neutral format for product data created and shared by different applications. Neutral means that the STEP data format will not favor one particular vendor.

IGES is an example of a neutral data exchange format [18]. IGES was originally intended to provide a means for exchanging engineering drawing data between CAD systems. One problem that occurred with IGES is an outgrowth of the way vendors implement the software that is required to translate their data to and from the neutral IGES data file. Currently, a vendor's translator can create IGES data files which contain data that makes sense in the context of their system. When that same IGES data file is loaded into another vendor's system, an incomplete data translation can occur because the second vendor's translator has made a different set of assumptions about the data it is receiving.

STEP goes beyond IGES both in the breadth of its information content and in the sophistication of its information system methodologies. In addition, STEP development is including the definition of subsets of product data that are specifically required for particular usage contexts. These subsets are called application protocols.

2. Application Protocols

STEP application protocols address the issues of completeness and unambiguity of data transfer by specifying in advance what data should be transferred in a particular context--thereby alleviating the need for vendors to make problematic assumptions. Application protocols are those parts of STEP that are relevant to a particular data-sharing scenario [27].

As explained in the previous section under "Commercialization of STEP," the development of application protocols permits the incremental implementation of STEP. There will be many STEP application protocols.

The concept of an application protocol allows vendors to build an application system that can interface with STEP data in a standard manner. In a sense, an application protocol is a standardized way of implementing a portion of STEP for a specific application. It is almost like a recipe for building an application [32]. The functional components are illustrated in Figure 8 and a flow diagram of steps in the development of an application protocol are shown in Figure 9.

The development of an application protocol involves incorporating specific application requirements into STEP, then testing the application protocol for completeness, correctness, compliance, and self-consistency. It is an iterative process [32].

Among the technical issues being resolved are:

- How application protocols will communicate with each other and the product data,
- Whether application protocols will be independent of the way in which the product data is used (for example, whether the data is shared or exchanged),
- Whether a commercial application must implement an entire application protocol or if it can utilize a subset of the application protocol, and
- How, and whether, information not already contained in STEP but needed by a new application protocol will be added to STEP.

The technical challenges involved in the development of application protocols are central to the use of STEP. Their development and implementation will determine whether STEP "can actually support complete, unambiguous exchange of product data across several application system boundaries" [32]. Application protocol development will force solutions to many of the remaining issues related to the usefulness and practicality of STEP itself.

3. Data Representations

At the core of the data sharing problem is data representation. STEP defines the information that describes products within different computer applications and across different enterprises. The use of computer software requires that the shared-data representations be specified. Data representation schemes must identify the data elements involved, their format, their meaning, and their relation

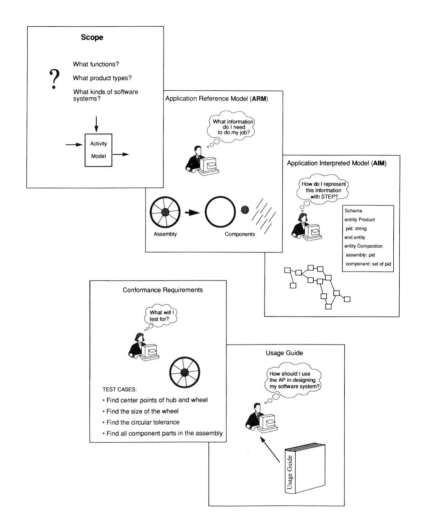

Figure 8. The five components of an application protocol are the scope, the ARM, the AIM, the conformance requirements, and the usage guide.

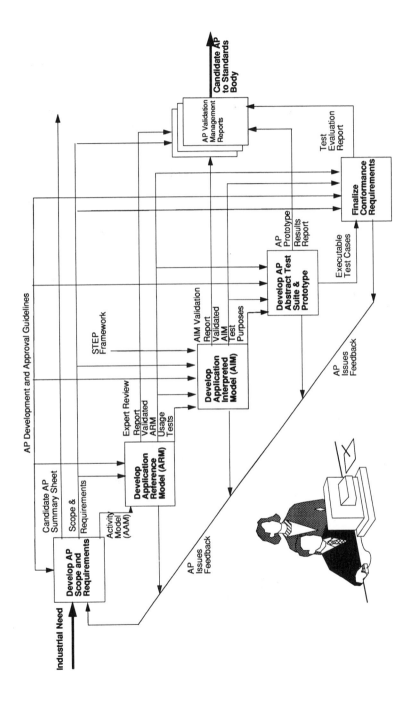

Figure 9. The steps in the development of an application protocol involve continual feedback to assure that the application protocol meets the needs.

to each other.   Data representations are formally defined within STEP specifications.

For example, in the geometry portion of the STEP specification, a simple data element may be called "point." The data representation for "point" might consist of three aspects: the point's X coordinate, its Y coordinate, and its Z coordinate. To complete the data representation, the type of numbers allowed for the point's X, Y, and Z coordinates must be explicitly stated.   In this case they would be "real" numbers, not integers or whole numbers.   Having defined the data representation for "point," other more complex data elements can also be defined that make use of the "point" data element.

Representations for data elements can become quite complex, making them difficult to define and understand. The most important criterion for the data representations used in STEP is that they must be unambiguous. This prevents their being misinterpreted by applications, or being interpreted differently by different applications.   Ambiguous data representations lead to problems like wires being mistaken for conduits, or bolts being mistaken for machine screws.

The developers of STEP employ information-modeling techniques to ensure that STEP will be unambiguous.   An information modeling language is actually used to define portions of the STEP specification. Implementations of STEP are written in EXPRESS [23].   EXPRESS has many features of a computer programming language. Writing STEP in EXPRESS allows information modeling experts to use specialized computer software to check the integrity, validity, and efficiency of STEP.   Besides facilitating the development of the standard itself, these information modeling techniques will also help to speed the development of future software applications based upon STEP.

STEP is organized into a framework composed of application information models, resource data models, and generic data models [33]. The generic data models, which integrate the resource data models, are the Generic Product Data Model and the Generic Enterprise Data Model [34]. The Generic Product Data Model (GPDM) contains information common to all products and meets the needs of application protocols by providing for the interpretation of generic facts in specific contexts [27]. The GPDM consists of the schemas: context, product definition, property definition, and shape representation.   Currently, the definitions of the schemas, in EXPRESS, are:

                    gpdm_context_schema
                        application_protocol
                        product_context
                        product_definition_context

                    gpdm_product_definition_schema
                        product
                        product_category
                        product_version
                        product_definition

product_definition_equivalence
product_definition_relationship

gpdm_property_definition_schema
  product_material
  product_shape
  shape_aspect
  surface_finish

gpdm_shape_representation_schema
  shape_model
  shape_model_composition
  shape_model_representation

The STEP data-sharing architecture must be able to access the data wherever and however it is stored. The data will be in a form dictated by the STEP Generic Product Data Model. A STEP data access interface may be the method used to provide application systems with the STEP data needed to perform an application [22].

4. Technical Evolution of STEP

A sound technical specification for STEP must address many issues pertaining to the architectures of information systems and the management of product life cycle data. Many different technologies have been brought together to establish a technical foundation for STEP. Computer-aided design and solid modeling systems provided the initial framework for describing product data. The fields of information modeling, relational and object-oriented database management systems have provided software tools that have contributed to the development effort. Technical experts who are familiar with the data requirements of design, process engineering, machine programming, and product support systems have helped define the types of data that must be supported in a product data exchange specification.

Because of the broad range of product types and application technologies which must be covered, the transformation of STEP from an abstract concept to a commercial reality is an evolutionary process. STEP application areas range from simple mechanical parts to complex electronics systems to buildings and ships. STEP is undergoing four stages of evolution:

Stage 1: Establishment of the foundation for STEP: The creation of a specification for the standard representation of product data involves many complex issues. It is virtually impossible for one individual or even a small group of individuals to write this kind of specification. The development of this specification requires both a strong technical and institutional foundation. The technical foundation for STEP is based upon a number of different information and manufacturing systems technologies and the experience of many technical experts. The institutional foundation is provided by voluntary technical activities, national and international standards organizations, businesses and industrial consortia, and government agencies. Because of the great need for consensus, all of these institutions must be in general agreement about the content of STEP, if it is going to be an effective standard.

Stage 2: Validation and standardization of technical specifications: Once an initial specification has been created, it must be validated, that is, tested to determine that it meets the needs of the user community. Validation testing takes into account how the specification will be used. Technical experts define the requirements for the different kinds of software applications that will use STEP and build information models based on the proposed STEP standards. These information models are then tested to determine whether they will meet the needs of state-of-the-art software applications. Test criteria, test procedures, and test data are also developed as part of the validation process. Only after satisfactory test results are achieved can the specification be considered workable and complete. The results and recommendations generated by validation testing flow back to the standards organizations for review and action.

Stage 3: Development of tools and prototype applications: The development of commercial STEP-based software products can be accelerated by prototyping. The developers of these prototype systems will learn a lot about using STEP technology that will help to accelerate the development of commercial products. The software tools that are developed may also be used in future products. If this work is done in the public domain, many companies can benefit from the results of this effort. Furthermore, early prototype applications can be used to validate the suitability of proposed standards. They can also be used for integration testing, that is, testing to determine whether or not different types of applications can work together. Prototype systems also may be extended to exercise conformance testing systems. In the absence of these prototype implementations, vendors and customers may make claims of conformance through self-testing.

Stage 4. Commercialization of and transition to STEP-based systems: Ultimately, STEP-based systems must be developed and marketed commercially. It will take a number of years for industry to recognize all of the different specialty niches for these systems and to develop stable products. Certainly this phenomenon can be seen in the personal computer market. Although the basic interface specifications for PCs were established in the early 1980's, new types of hardware and software products are still being defined today. It will undoubtedly take a number of years after products become available until they are put into widespread use within industry and government. Considerable advanced planning and investment of resources will be required to transform large government and industrial organizations into new STEP-based systems. Translation planning is essential to implementation and acceptance of STEP.

The first stage of STEP evolution is well underway, but the second stage has just barely started. Stages 2 through 4 will also have to be repeated for the different product technologies that STEP must cover, such as mechanical assemblies, sheet metal parts, structural systems, and electronics components.

5. Verification and Validation

Verification and validation are two ways in which commercialization of STEP-based products can be expedited. *Verification* is the review of both the system requirements to ensure that the right problem is being solved and the system design to see that it meets those requirements. *Validation* is the test and evaluation of the integrated system to determine compliance with the functional, performance and interface requirements that were verified. Validation and verification are necessary during the development of STEP and its associated software tools, as well as for the development of STEP-based products, that is, STEP implementations.

With respect to the development of STEP, validation requires testing to confirm that the requirements for the product life cycle data have been met. One of the major goals of the validation testing efforts is to test the suitability of the proposed STEP standard for product life cycle information systems applications.

Validation testing is aimed at evaluating the completeness and the integrity of the STEP specifications. Without validation testing, many deficiencies in the specifications might not be discovered until commercial applications are constructed. It is obvious that without this testing, developers might have had to bear the burden of excessive redevelopment costs and delays while the specifications are "fixed."

Validation testing is discussed more thoroughly in Section E, The NIST National PDES Testbed.

6. Application Systems

Application systems are the computer software systems that will use STEP. They are systems for computer-based manufacturing functions such as computer-aided design, analysis, manufacturing planning, resource allocation and scheduling, manufacturing equipment programming, and quality assurance. Many of these systems have common data requirements and they need to share data. (A simple example of shared data is the name of the product and the identifiers of its component parts.) The early development of prototype STEP application systems is the key way to accelerate commercialization of STEP.

Some product data requirements may be unique to a specific type of application. For example, the tolerances on a product's dimensions would be required by manufacturing planning systems, but this same data would be irrelevant to scheduling systems. Yet both systems would refer to the same names when identifying the product and its components.

Ensuring that STEP addresses the requirements for manufacturing applications is a significant challenge. (This was discussed in the context of application protocols.) Generally, there are no formal, publicly available specifications of the information requirements for any of these systems. Functional requirements and design specifications must be developed for systems that will use STEP. These specifications should be defined concurrently with the evolving application protocols. They will help to determine exactly how STEP will be used by future commercial systems.

Prototype application systems should be developed that can be used to test the viability of the application protocols. Different types of prototype systems should be tested with each other to ensure that STEP permits interoperability between various applications. If the prototypes are constructed in the public domain, they can later be used as foundations and building blocks for commercial implementations.

7. Configuration Management

The process of developing an information processing standard involves the creation and management of thousands of documents and computer programs. Knowing which documents and computer programs are current and which are obsolete is critical to the development process. Configuration management provides the fundamental operational capability for tracking and maintaining versions of documents and software.

Configuration is the logical grouping and/or collection of elements into a coherent unit. This unit is typically a version of a software release or text document. If the configuration of an information unit is to be controlled, access and changes to the information must be controlled. Often "master" documents and approval mechanisms are established to ensure the quality and integrity of the information that is being managed.

The complexity of the configuration management problem is governed by the type of information involved and how it is to be controlled. In the case of simple configuration control systems, for example those that deal with software source code control, simple text files are usually just grouped together into a named or numbered unit and distributed as a single item. This is a simple process and many software products currently perform just this function. The complexity of the problem increases when the configuration involves more than just simple text files. Two examples of more complicated configuration control problems are the management of computer programs which run on different computer systems, and documents which include graphic images.

Clearly, the development of STEP is a complex configuration management problem. It involves a number of different organizations that have different interests in the technical aspects and in the status of the proposed standard. Each organization must be able to retrieve proper versions of the developing standard. Software tools are needed which can be used to merge electronic versions of text and produce a single unified document from each organization's contributions. This assembly process is one of the main functions of a good configuration management system. Reliable, controlled, and up-to-date access to an individual organization's data plus the capability to pull disparate pieces of information together is a major challenge.

The discussion of configuration management is continued in Section E., The NIST National PDES Testbed.

8. Conformance Testing

Before commercially developed systems are marketed, conformance testing procedures must be established which act as quality assurance mechanisms to protect both system developers and users. Conformance testing is the evaluation process or methodology that is used to assess whether products adhere to standards or technical specifications. If independent conformance testing mechanisms are not established, customers will have to accept vendor assurances that their systems comply with STEP. Unfortunately, many vendors may be incapable of determining whether or not their products faithfully comply with the standard.

The development of conformance testing methods and the development of application protocols are intertwined. Commercial systems based upon one or more application protocols will be the first implementations of STEP. Conformance testing methods for STEP will only be based on evaluating implementations of application protocols.

D. The Role of NIST: An Engineering Paradigm

Research and hands-on experience are essential for NIST scientists and engineers to make informed and impartial standards recommendations. Recognizing the importance of manufacturing interface standards, NIST established the Automated Manufacturing Research Facility (AMRF) in 1980 to investigate critical issues in factory automation standards. The first major goal of the AMRF involved the construction of a flexible manufacturing system, a testbed, for the small-batch manufacturing environment.

The facility is used as a laboratory by government, industry, and academic researchers to develop, test and evaluate potential interface standards. To ensure that the interface standards issue is addressed, the testbed is designed to contain component modules from a variety of vendors [35].

The AMRF represents a fresh approach to factory automation. The generic factory architecture incorporates elements such as hierarchical facility control, distributed database management, communication network protocols, on-line process control (deterministic metrology), data-driven (and feature-driven) processes, and manufacturing data preparation (such as design, process planning, and off-line equipment programming).

It is this experience in building a large-scale testbed facility, working with industry and universities, studying standards issues, and implementing testbed solutions that brings NIST to an important role in the development and implementation of PDES/STEP.

1. Components of the NIST Engineering Paradigm

Traditionally, engineering projects have been carried out by starting with specifications, developing or adapting technology, and developing the required application. Today, the management of information--especially information in electronic form--has become a critical component of any engineering endeavor. This is especially true in the work of the NIST Factory Automation Systems Division, where much of the PDES/STEP work at NIST is done. The paradigm can be used as a model for understanding and planning engineering projects.

The paradigm consists of four components:

- System Specification,
- Information Management Technology,
- Engineering Technology, and
- Engineering Application.

The paradigm is shown diagrammatically in Figure 10. The system specification component takes an industrial need such as "manufacturing world-class products" and develops the information and functional models that address the needs. The information management technology component takes the standards, in this case product and manufacturing data standards, and generates the information framework or architectecture concepts required to implement an engineering application. The engineering technology component takes the functional requirements for the applications as determined by the system specification and creates the engineering framework or architecture concepts required to implement the engineering application. Finally, the engineering application component is the integration of the two technology components into a prototype application environment to test fully the proposed set of standards. The experience gained in the application environment is used to strengthen the system specification component. The outputs are indicated by the double-lined arrows: a set of standards from the system specification component and products from the engineering application component.

The combination of the need for advances in concurrent engineering technologies and the need to represent engineering data in a standard format--

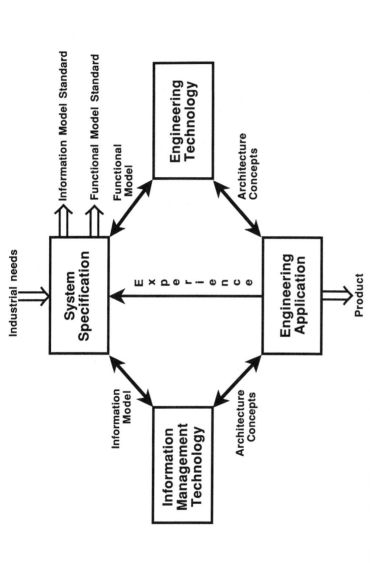

Figure 10. The relationships among the four components of the engineering paradigm involve models and concepts that carry out the transfer of information among the components. The double-lined arrows represent inputs to and outputs from the system (for example, needs, standards and products.

STEP--is a perfect industrial problem to be implemented using the engineering paradigm. This is illustrated in Figure 11 and by the following discussion.

2. System Specification

The function of the first paradigm component, system specification, is to take industrial needs and develop the information and functional specifications required to solve them. These specifications become the basis for the development of the information and engineering technologies required to implement a solution to the industrial needs. Most of the activities performed in this component are involved with the voluntary national and international standards programs.

Therefore, in the paradigm as applied to product data-driven engineering, system specification is the development of the STEP standard as an ISO data exchange standard and the implementation of application protocols that specify the engineering environment in which STEP is to be used. NIST participates in both the formal standards organization and in the research and development of testing procedures for STEP. A staff member serves as chair of the volunteer IGES/PDES Organization. Staff also actively participate in the technical committees within the IGES/PDES Organization where robust information models that define the scope and application of STEP are developed.

NIST scientists are involved in applying to STEP the Information Resource Dictionary System (IRDS) standard being developed by ANSI [36]. There is a project that addresses the application of IRDS to STEP, including using an IRDS extendibility feature to support the storage and management of the diverse conceptual and data models of STEP. In addition, work is going on to extend the STEP information resource dictionary schema to support a full three-schema architecture, to interface STEP IRDS to software such as conceptual modeling tools and database management systems, and to develop relationships to physical design for STEP.

NIST staff are involved also in identifying the application of geometric modeling to the definition of STEP and its application implementations. There are many technical issues, such as:

- *Interaction between different modeling geometry systems.* As an example, for NURBS (Non Uniform Rational B-Spline) surfaces, what is the best transformation from a 5th-order curve to a series of 3rd-order curves. In general, how is geometric information exchanged between constructive solid geometry, boundary, and wireframe models.

- *Topology and its relationships to geometry.*

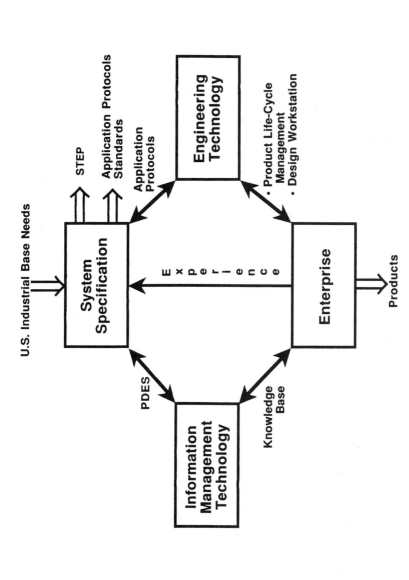

Figure 11. When the engineering paradigm is applied to concurrent engineering, the application laboratory is the prototype enterprise.

- *Geometry and topology and their relationship to application areas such as numerical control (N/C) coding, graphics display, collision detection, etc.*

An overriding issue is the problem of deciding what type of geometric modeler is appropriate for a given application. Research into how to categorize modeler parameters and measure expected performance for applications such as inspection and N/C coding is also important.

NIST staff participate in the development of testing procedures for STEP as well as STEP-based industrial products. They have developed test plans that identify the approaches, methodology, resources, and tasks required to test and validate STEP. There are many common testing methods that will be explored including, test data file, syntax analysis, semantic analysis, instance tables, and ad hoc database queries.

Testing of the specification and implementations are performed at the National PDES Testbed, described in the next section. The testbed will become a model for a network of future testbeds that will be established throughout the world. The Testbed will also serve as a model for the type of software and hardware configurations and personnel resources needed to test and implement STEP.

3. Information Management Technology

The function of the information management technology component is to develop the proper technology to process the information identified in the system specification. The important point to be stressed is that the technology (file system or relational database, for example) must be appropriate to meet the needs of the engineering technology. The following are typical tasks to be performed:

- *Determine from the information model specification the types of data representations required.* Areas of concern include the implementation of a data dictionary, the types of schemas for representing the informational relationships, and the extent to which knowledge (rather than just information) needs to be represented.

- *Based on the engineering environment, such as flexible manufacturing system or robot control, determine the important characteristics of the information management technology needed.* The issues may include, for example, distributed vs. central storage, homogeneous vs. heterogeneous computing, version control (or configuration management), time constraints, database size, and security.

- *Design and implement an information management system capable of handling the required characteristics.*

In short, this component of the engineering paradigm is concerned with the conversion of STEP into an information management system that can support the engineering requirements.

4. Engineering Technology

The function of the engineering technology component of the paradigm is to convert the functional specifications into a collection of engineering concepts and a systems architecture that can address the industrial problem. The following are typical tasks to be performed:

- *Define a plan for developing the technology.* This includes decomposing the overall problem into a series of tasks and specifications.

- *Identify the product data requirements and measurement systems needed.* Define the control architecture and process interfaces. Develop new engineering concepts to address the problem.

- *Design the overall system, including the information and functional models.* Define the data requirements and the means by which data is to be collected and analyzed.

- *Determine which processes are needed for the given application, for example, process planning or robot handling.*

5. Engineering Application

The fourth component, engineering application, is the prototyping of the concepts and architecture defined in the two components, information management technology and engineering technology. The outputs of the engineering application component are feasibility demonstrations of how the engineering and information management concepts result in a credible solution. The following are typical tasks that are performed:

- *Based on the systems specification and the technology to be developed, specify the product mix to be used in the laboratory.*

- *Develop the interfaces between the information management systems and the engineering processes.* Develop the interfaces between the various processes that compose the engineering application.

- *Build a laboratory based on the architecture and concepts defined for the information and engineering technologies.* Design and perform experiments that provide the proof-of-concept for the technologies.

The engineering application paradigm component is realized as a laboratory in which the architecture and concepts developed in the information management and engineering technologies components are implemented. Facilities that represent processes that are part of the product life cycle are built and experiments are conducted to test the technology concepts.

At present, the AMRF can be viewed as the engineering application for the subset of the product life cycle that addresses the design, manufacturing and inspection processes. The work in manufacturing data preparation, process control, and factory control addresses the engineering technologies for flexible manufacturing. There are also efforts in data management and network communications that address the information management technology issues.

The factory control system for the manufacturing and inspection of parts uses a STEP-like format. The systems are all data driven. In fact, the vertical workstation is driven from an off-line programming environment that starts from a set of machinable features for a part [37]. The inspection workstation is driven from an off-line programming environment that generates a CAD database of the part with respect to the necessary tolerance information [38]. The five-level control architecture developed within the AMRF has become a model for the implementation of manufacturing systems.

A common thread throughout the AMRF is the standardized method of handling data. This is particularly true in the manufacturing data preparation research which is aimed at a seamless architecture based upon plug-compatible modules that streamline the preparation of data for automated manufacturing systems.

In the AMRF, incoming part descriptions are converted to AMRF Part Model Files using commercial CAD systems and software developed for the AMRF [39]. The AMRF Part Model File includes 3-D geometric and topological information, tolerances, and other data on the part in a uniform format that can be used by other AMRF systems. Translators have been written to convert this format to STEP.

Working from the STEP files, and other information in the database system, operators then prepare "process plans" for the part. In the AMRF, these computerized plans include the cell's "routing slip," which is used to schedule the movement of materials and the assignment of workstations; the workstation "operation sheets," which detail the necessary tools, materials, fixtures, and sequences of events; and the machine tool's "instruction set," which guides the tool through the motions required to shape the part. Research is being conducted into the development and testing of a single set of standard data formats for process planning at every level of the factory control hierarchy, and an editing system to generate, archive and update these plans. In effect, the AMRF provides a laboratory for a STEP implementation. The AMRF approach to handling data is to allow the users freedom to select computers and database software, yet still be able to build an "integrated" system.

Ideally, a factory control or planning system should be able to request the information it needs without knowing which of several databases holds the information, or what format is used to store the data. A distributed database management system called the Integrated Manufacturing Data Administration System (IMDAS) is used in the AMRF to meet this need [40].

The AMRF data communications system allows computer processes such as control programs to run on many different computers and to be developed using different languages and operating systems. This system uses a method of transferring information through the use of computer "Mailboxes," which are areas of shared memory on various computers to which all machines have access through the network communications system [41].

The Manufacturing Systems Integration Project uses the AMRF as a testbed to study data and interface requirements for commercial manufacturing engineering software systems. The concentration is on the modeling of and access to information needed by manufacturing systems during production and the integration of those systems, such as scheduling and control as shown schematically in Figure 12. The goals include demonstrating feasibility and testing integration and interface concepts for information standards to integrate manufacturing engineering and production systems.

In essence, the AMRF is the laboratory where control and metrology concepts and architectures for integrating information and technologies are implemented and tested.

Other laboratories in the Factory Automation Systems Division fulfill the paradigm expectations and perform a function similar to the AMRF for specific application areas. They include the Engineering Design Laboratory [42], which is used to evaluate software tools for integrating design and analysis and for modeling design intent and design knowledge for access and use throughout the life cycle of a product. Another example is the Apparel Design Research System, which is used to help develop methods for product data exchange that are appropriate to the apparel industry [43]. (The design project is funded in part by the Defense Advanced Research Projects Agency and the apparel project is funded by the Defense Logistics Agency.)

E. The NIST National PDES Testbed

The NIST National PDES Testbed is a focus for planning, coordination, and technical guidance of a national effort for STEP development and implementation. The national effort consists of a growing network of participating organizations of various types.

Located at the National Institute of Standards and Technology, the Testbed is a publicly accessible facility where the STEP specification and STEP-related tools can be modeled, analyzed, prototyped, implemented, and tested [44]. Physically,

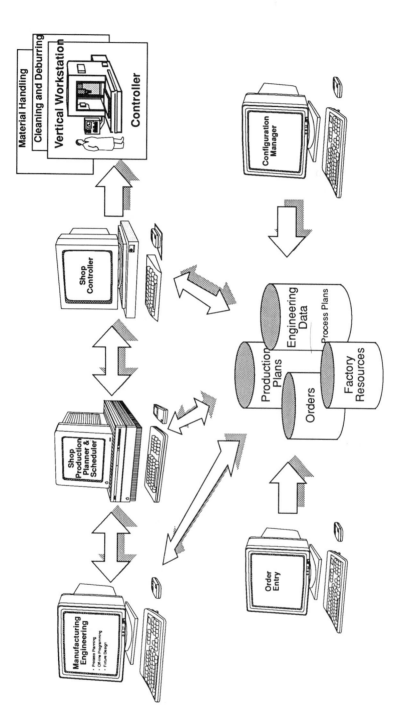

Figure 12. The Manufacturing Systems Integration Project builds upon previous AMRF work in hierarchical control architectures, the representation of manufacturing data, a language for process specification, and the integrated manufacturing data administration system (IMDAS).

the facility is comprised of laboratories, computer hardware and software systems, and testing tools. The laboratories include unique laboratories such as the Validation Testing System, as well as multipurpose laboratories such as the AMRF and the Engineering Design Laboratory. The Testbed is used and staffed by leading experts on PDES issues from industry, academia, and government. It is currently staffed with the full-time equivalent of approximately 20 scientists, engineers, and support personnel.

The National PDES Testbed supports the goals of the IPO and ISO to establish an international standard for product data sharing. The Testbed was established at NIST in 1988 under U.S. Department of Defense Computer-aided Acquisition and Logistic Support (CALS) program funding. Standards which support product data sharing are recognized as a major building block in the CALS program. Under CALS sponsorship, the National PDES Testbed is advancing the development of product-sharing technologies. The staff of the National PDES Testbed are not only involved with the ISO and IPO, but also actively participate in the technical activities of PDES, Inc.

The Testbed is also the cornerstone of the Manufacturing Data Interface Standards Program at NIST. The goal of this program is the development of national standards for a "paperless" manufacturing and logistic support system.

The overall objective of the Testbed is:

To provide technical leadership and a testing-based foundation for the rapid and complete development of the STEP specification.

The major functions of the Testbed include:

Standards validation test development to ensure that the specifications and underlying information models meet the needs of product life cycle systems;

STEP application prototyping and interoperability testing to provide test cases, tools for generating test cases, and application experts who can critically evaluate the draft specifications; to ensure that the specifications are sufficiently integrated to guarantee interoperability of different types of STEP applications; and to demonstrate the advantages and suitability of STEP for use in industrial environments;

Product data exchange network integration to provide a national network at government and industry manufacturing sites and laboratories to share information and test cases; and

Configuration management to implement a configuration management system and establish a central repository for documents and software generated by various organizations involved in the STEP development process.

1. Standards Validation and Conformance Testing

Validation testing is the process that ensures that STEP is usable and functional. It confirms that the standard is complete, unambiguous, and consistent. It determines that the standard meets the needs of the user community. The results and recommendations generated by validation testing must be fed back to the standards organizations for review and action. Standards committee members may then amend the specifications, affected portions may be re-tested, and the specifications can be approved as standards.

The emphasis on validation at the Testbed is on the development of computer-assisted tools for testing and evaluating proposed application protocol specifications [28].

The validation process is evolving along with STEP itself. Technical challenges still remain, including such issues as the degree of functionality that must be defined in an application protocol and that must be achieved by application systems.

To support validation testing, the Testbed provides an integrated computing environment. In addition, it acts as a repository for proof of the qualities that the STEP specification exhibits. This proof, in the form of test results and real-world test product data, will help the standardization process to proceed and will encourage implementations of information systems which use STEP.

The Validation Testing System within the Testbed is comprised of software that will: 1) automate the evaluation of the computable qualities, such as whether or not the syntax of the specification language was followed, and 2) assist validation teams with solving intuitive problems which are not feasible to automate. The names of the major component modules of the validation testing system are:

- Model Scoping and Construction Tool
- Test Definition Tool
- Test Case Data Generation Tool
- Test Case Execution and Evaluation Tool

Figure 13 illustrates the major validation testing tools and their functions. Just as validation testing is essential to the development of STEP, conformance testing is essential to its successful implementation. Conformance testing is the testing of a candidate product's behavior and capabilities. The behavior and capabilities of the product must be those required by the standard itself, and they must be exactly what is claimed by the manufacturer of the product.

Conformance testing helps to assure product conformity in implementations, clarifies the standard itself for implementation, provides a feedback loop to the standards-making bodies for improvements to the standard, and encourages commercial development by providing a baseline for commonality in all products. It does not guarantee that the product conforms to the standard, nor does it assure that the product is of high quality or reliability.

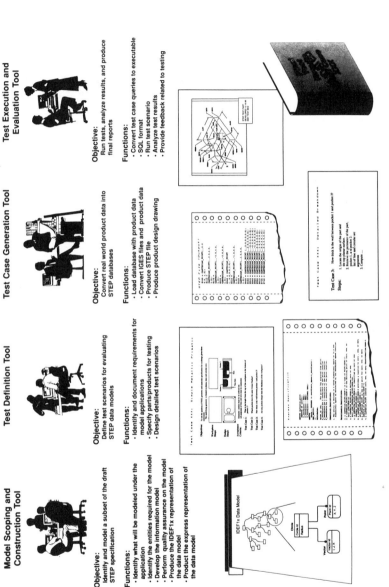

Model Scoping and Construction Tool

Objective:
Identify and model a subset of the draft STEP specification

Functions:
- Identify what will be modeled under the application
- Identify the entities required for the model
- Develop the information model
- Perform quality assurance on the model
- Produce the IDEF1x representation of the data model
- Product the express representation of the data model

Test Definition Tool

Objective:
Define test scenarios for evaluating STEP data models

Functions:
- Identify and document requirements for model applications
- Specify parts/products for testing
- Design detailed test scenarios

Test Case Generation Tool

Objective:
Convert real world product data into STEP databases

Functions:
- Load database with product data
- Convert IGES files and product data
- Produce STEP file
- Produce product design drawing

Test Execution and Evaluation Tool

Objective:
Run tests, analyze results, and produce final reports

Functions:
- Convert test case queries to executable SQL format
- Run test scenario
- Analyze test results
- Provide feedback related to testing

Figure 13. **The validation tools used in the National PDES Testbed include model scoping and construction, test definition, test case generation, and test execution and evaluation.**

The implementation of a conformance testing system and an independent testing program increases the probability that different STEP implementations will be able to interoperate. Figure 14 shows the conformance testing program model.

The National PDES Testbed will construct a conformance testing system [45]. In cooperation with others, the Testbed plans to develop test procedures and data that adhere to STEP application protocols, specify the process which will be used for certifying compliance with the standard, and define the procedure which will be used to approve and review the operations of testing laboratories. The Testbed intends to help establish a conformance testing program at selected sites on the Product Data Exchange Network.

The standardization and acceptance of a conformance testing methodology, as well as appropriate test methods will allow producers to test their own products through a testing laboratory and will lead to acceptance of test results from different testing laboratories.

2. Application Prototyping and Interoperability Testing

For application prototyping and interoperability testing, the Testbed includes a "STEP Production Cell." The STEP Production Cell will demonstrate small batch manufacturing using STEP data [46]. It will be an integrated, automated manufacturing environment within the NIST AMRF whose product specification data representation is based upon validated STEP data models. It will help verify that the STEP standard is workable through production level testing. In cooperation with test sites having similar capabilities, the STEP Production Cell will test and demonstrate how STEP supports production operations occurring at different sites.

The STEP Production Cell will integrate basic STEP software tools, commercial databases, and commercial manufacturing applications into a prototype small-scale manufacturing environment. Within this environment, it will be possible to verify the performance of STEP under real-world conditions and to demonstrate STEP-based manufacturing across different production sites.

The manufacturing data preparation subsystems of the STEP Production Cell are design, process planning, and equipment programming. These subsystems are used to generate the information that is required to control the manufacture and inspection of a part. STEP data is the primary information shared by these subsystems.

Within the cell, the Machining Workstation is a 3-axis vertical milling machine. This computer-driven machine tool can produce simple, prismatic parts. The computer programs that control this machine tool are derived from the STEP data provided by the manufacturing data preparation subsystems.

The inspection workstation, a coordinate measuring machine, provides the facility for determining whether machined parts are produced as specified. Based on measurements from the coordinate measuring machine, analysis software determines whether dimensions of the machined part fall within designed

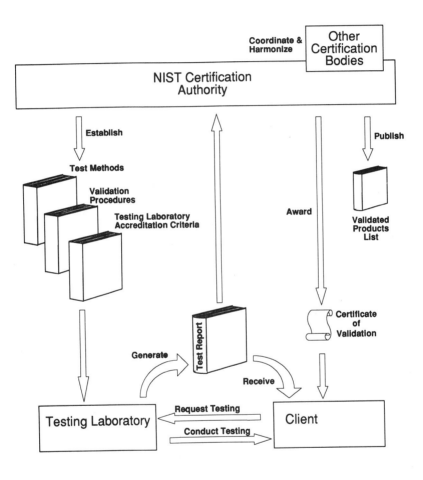

Figure 14.   **Conformance testing and certification is depicted in the conformance testing process model.**

tolerances. As with the milling machine, the computer programs that control the measurement process are derived from the STEP data provided by the manufacturing data preparation subsystems.

The data repository subsystem provides the storage mechanism for STEP data. The repository provides a generic software interface to the data representations. The generic interface allows the application subsystems to store and retrieve the desired STEP data without regard to the details of its representation. The network communications subsystem ties the other six subsystems together.

Figure 15 describes some of the major processes and information contained within the STEP Production Cell.

3. Product Data Exchange Network Integration

The Product Data Exchange Network will be a network of organizations and individuals dedicated to support the specification, validation, prototyping, commercial development, and conversion to STEP. The Network will help accelerate the realization of STEP and will help ensure that STEP will function as intended in actual manufacturing environments [47].

The Network will consist of a broad spectrum of manufacturing facilities and research centers from industry, academia, and government linked electronically via computer networks. The plan is to begin with the AMRF-based experience in mechanical parts, then to expand into other areas. Eventually, the Network will include sites in various manufacturing domains, such as aerospace, shipbuilding, apparel, sheet metal products, electrical products, and others. A goal of the Product Data Exchange Network is to accelerate the transition of these facilities to STEP-based information systems.

The National PDES Testbed will serve as headquarters for the Product Data Exchange Network. Because the Network and the CALS Test Network sponsored by the U.S. Department of Defense have similar objectives, the activities and results of these two programs will enhance and complement each other.

Several of the network sites will serve as model facilities for developing STEP-based manufacturing systems. Various Product Data Exchange Network sites will perform STEP validation activities based upon specific capabilities available at that site. These activities may include testing or developing STEP-based software applications, developing transition plans to implement STEP in manufacturing environments, or producing actual products using STEP. Figure 16 depicts some of the activities which may occur at Network sites.

4. Configuration Management

The National PDES Testbed provides configuration management systems and services for key organizations participating in major PDES and STEP activities. The Testbed configuration management system can be used to control access and distribution of documents and software. In the future, product models and

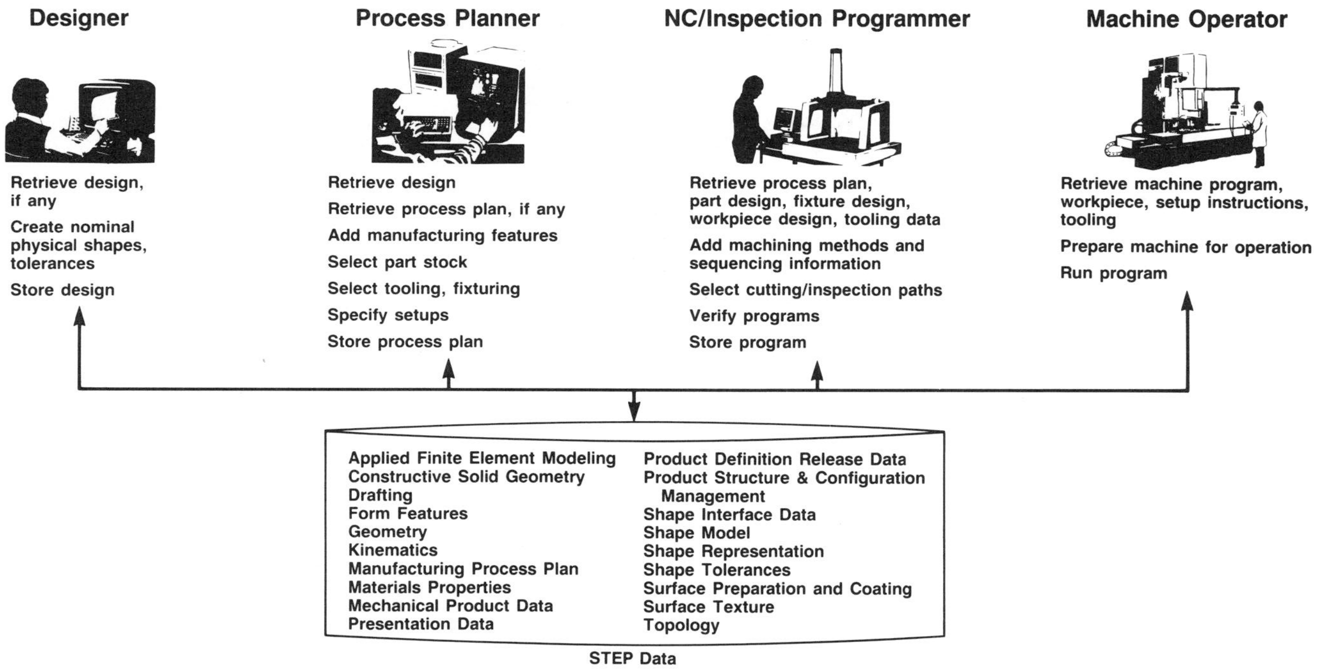

Designer

Retrieve design, if any
Create nominal physical shapes, tolerances
Store design

Process Planner

Retrieve design
Retrieve process plan, if any
Add manufacturing features
Select part stock
Select tooling, fixturing
Specify setups
Store process plan

NC/Inspection Programmer

Retrieve process plan, part design, fixture design, workpiece design, tooling data
Add machining methods and sequencing information
Select cutting/inspection paths
Verify programs
Store program

Machine Operator

Retrieve machine program, workpiece, setup instructions, tooling
Prepare machine for operation
Run program

Applied Finite Element Modeling
Constructive Solid Geometry
Drafting
Form Features
Geometry
Kinematics
Manufacturing Process Plan
Materials Properties
Mechanical Product Data
Presentation Data

Product Definition Release Data
Product Structure & Configuration Management
Shape Interface Data
Shape Model
Shape Representation
Shape Tolerances
Surface Preparation and Coating
Surface Texture
Topology

STEP Data

Figure 15. The processes and information within the STEP Production Cell of the National PDES Testbed will help verify the workability of the standard in a prototype manufacturing environment.

Figure 16. The Product Data Exchange Network will allow a variety of organizations to participate in the development of STEP.

graphical representations will be included [48]. The functional architecture of the system is shown in Figure 17.

The core of the configuration management system is based upon a general set of common requirements. Customized interfaces will be constructed which account for each organization's internal processes and procedures.

F. The Extension of an Enterprise Integration Framework

Concurrent engineering is an engineering approach that can help optimize the operations of a manufacturing enterprise. However, the optimization is "localized" to the life cycle--design to production to support--of the enterprise's product. Clearly, concurrent engineering is but one dimension of a bigger idea. That bigger idea is *the optimization of all the enterprise's operations, including planning, marketing, and financial operations, as well as its transactions with its suppliers, distributors, and other business partners.* "Multi-enterprise concurrent engineering" is the term that connotes the broader optimization. This broader optimization is based upon the integration of all the operations within an enterprise and between an enterprise and its business partners.

The term for the standard architecture that would allow the integration of all activities of manufacturing enterprises is "enterprise integration framework." Just as STEP implies a standard means of representing information about a product *as well as* the infrastructure necessary to access and contribute to that information in a heterogeneous computer environment,

> *Enterprise Integration Framework* includes the structure, methodologies, and standards to accomplish the integration of all activities of an enterprise.

The key is the sharing of all kinds of information that allows for a concurrent approach not only to engineering, but also to accounting, marketing, management, inventory control, payroll, and other activities that are vital to the functioning of an enterprise. Multi-enterprise concurrent engineering through an enterprise integration framework is an approach that can both *guide the integration* of an enterprise's activities and *provide the standardized organization and arrangement* for the integration to occur.

Just as in the implementation of computer integrated manufacturing (CIM) [49], the major technical challenge to an enterprise integrated framework is the design of the integrated system architecture. Beginning with a system architecture, developing the methods to build the models, and then building an integrated framework is the "top down" approach to the integration of all components of an enterprise. A number of architectures have been proposed for CIM [50], but enterprise integration architectures have been studied only recently.

Because every company is unique in the way that it operates and because there are different laws and cultures in different countries that affect how businesses

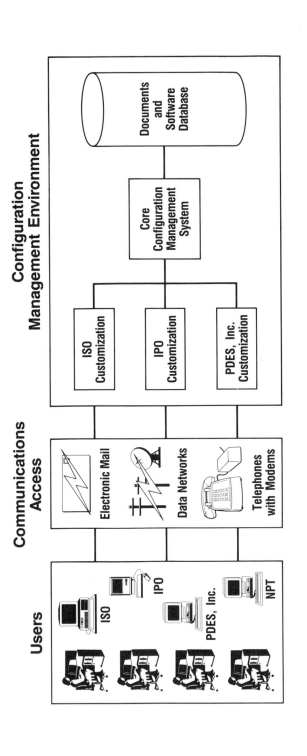

Figure 17. The Testbed configuration management system will provide convenient access to STEP-related documents and software for participants in ISO, the IGES/PDES Organization, and PDES, Inc., as well as the National PDES Testbed (NPT).

operate, it is essential that an enterprise integration framework be flexible and conceptually broad. This is the realm of enterprise modeling. [51]

Enterprise modeling is the abstract representation, description and definition of the structure, processes, information, and resources of an identifiable business, government activity, or other large entity. The goal of enterprise modeling is to achieve model-driven enterprise integration and operation. Also important are modeling techniques for describing the logistic supply chains in an industry, including the business processes that occur among independent but closely cooperating enterprises.

It is also essential that such a large undertaking as enterprise integration framework development be carried out internationally. The consensus development of international standards for integrating enterprises will help assure that the benefits of concurrent engineering approaches, as well as opportunities for global economic competitiveness, are available to all enterprises.

Open Systems Architecture (OSA) is the description of those computing and networking systems that are based on international and *de facto* public domain standards, rather than the proprietary systems dominating the current business environment. The concept is to be able to create modular information technology components, thus providing for a "plug and play" ability to swap out both hardware and software components among various vendor products. Complex products for OSA require substantial investment and development time. Much of the OSA product planning is precompetitive and linked to standards activities that require coordination.

In the U.S., a number of government agencies are initiating efforts to define and develop an enterprise integration framework. These agencies include the Air Force (through the Wright Research and Development Center's Manufacturing Technology Directorate), the Defense Advanced Research Projects Agency (DARPA), the CALS office under the Office of the Secretary of Defense, and NIST. A major goal is a set of international standards that provide a framework upon which commercial (and government funded) information technology related products could be produced that will support multi-enterprise information systems for industrial applications.

The Air Force Enterprise Integration Framework Program is intended to provide a common reference model for establishing research priorities, harmonizing standards development efforts, and developing a strategy for coordinated investment by government and industry in automated infrastructures. It is anticipated that an international consensus can be built for use of this framework as the model for the development or implementation of international standards and for integrating many types of applications and industries. The program is part of the U.S. effort to cooperate internationally in a coordinated program to define, develop, and validate a conceptual framework for inter- and intra-enterprise integration based on open systems principles and international standards.

Within the European Strategic Program for Research on Information Technology (ESPRIT), a government-industry European CIM Architecture

(AMICE) consortium is working to develop a Computer Integrated Manufacturing Open Systems Architecture (CIM OSA) [50], [52].

In a sense, just as multi-enterprise concurrent engineering is the next step in the evolution of manufacturing, the enterprise integration framework is the next step in the evolution of engineering standards. As indicated in Figure 18, engineering education will have to evolve also. Perhaps product data engineering will become as important as the traditional engineering specialties were in the early part of this century.

A vision of the future manufacturing environment is shown in Figure 19. Independent enterprises operating as suppliers, system integrators, merchants and customers are integrated by an information framework into an effective system. Within each of these enterprises, the various product-related functions and product life cycle stages are integrated through the sharing of product data, although each stage maintains its own view of the product. Based upon standards, the inter- and intra-enterprise integration enables the practice of multi-enterprise concurrent engineering [53]. It is the practice of multi-enterprise concurrent engineering through which the characteristics of world-class products are achieved. These characteristics are short-time-to-market, low cost, high quality, and high functionality.

VI. CONCLUSIONS

The primary aim of any manufacturing enterprise is to deliver working products to customers. To this could be added timeliness, cost effectiveness, quality, reliability, and other characteristics that contribute to a competitive product and hence to profits. Nevertheless, the bottom line is simply *working products in the hands of satisfied customers*.

Recently there has been increased recognition that concurrent engineering, engineering design, manufacturing engineering practices, and data exchange and interface standards are critical to international competitiveness [54] [55]. These technologies, based upon information technology in general, are the means for providing to customers high quality and reliable products, as well as the support for those products, in a timely and cost-effective manner.

Information technology will provide an integrated level of automation based upon standards and frameworks. It will create a climate in industry in which enterprises can benefit from cooperation, collaboration and interdependence, without sacrificing their individual independence, initiative, and intellectual property rights. Information technology, by enabling such approaches as concurrent engineering, will stimulate the necessary standardization and provide the economies of scale that would not be otherwise provided without drastic changes in the way businesses in the U.S. operate.

Concurrent engineering, based upon information technology, will initiate a new industrial revolution. Certainly, the bottom line would still be *working products in the hands of customers*, but future products are much more likely to be of higher

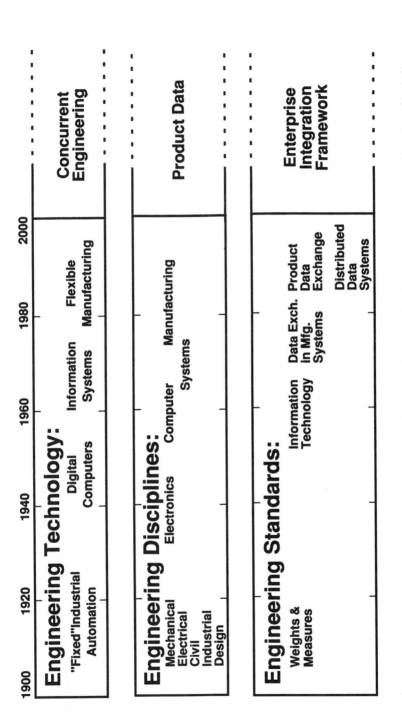

Figure 18. As the evolution of manufacturing and engineering standards progresses, new engineering disciplines are required. "Product data engineering" may become the next new engineering discipline.

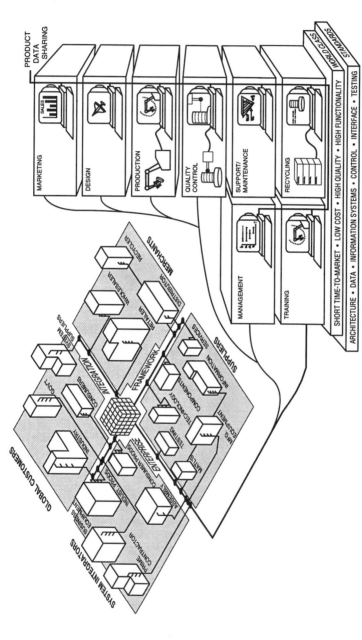

Figure 19. Through an enterprise integration framework (indicated by the symbolic rectangular framework) enterprises will be integrated into a system that encourages the practice of multi-enterprise concurrent engineering. The structure at the right is an exploded view of one representative manufacturing enterprise that shows how the various internal departments in each enterprise will be integrated through the sharing of product data. Product data standards will allow each functional department to retain its own view of a product.

quality and more reliable, state-of-the-art products at prices that are much lower than they might have been if concurrent engineering were not used.

It is instructive to reflect on the mechanical drawing and the way it impacted the entire manufacturing process in its era. Prior to the industrial revolution, manufacturing was defined by a physical model of a product to be reproduced. For example, a worker would ensure that the dimensions of the product to be produced corresponded to the model by using calipers to transfer measurements from one to the other. This method reinforced the tradition that workers manufactured complete but specific product types rather than generic components of products.

In 1801, Gaspard Monge wrote "La Geometrie Descriptive." It was the first treatise on modern engineering drawings. It described the concept of projecting dimensioned geometric views of an object onto three perpendicular planes. Since it included size and shape information, the mechanical drawing became an objective standard of performance for workmanship and thus the need for a model was eliminated.

The drawing enabled the practice of designing a product with interchangeable parts. A product could be produced by contractors who could manufacture different components to be assembled. This capability led to the fragmentation of the manufacturing process that exists to this day. Moreover, in today's industrial enterprises, the life cycle processes for a product are no longer even performed by the same group of people. In fact, the processes are distributed through a network of factories.

The mechanical drawing concept has lasted for almost 200 years. Although it is a method for describing products, just as the physical model had been, the mechanical drawing revolutionized the manufacturing process itself. The drawing became the output of the design phase of the process and the input into the production phase. Drawings were converted into production process plans which were converted into programs or procedures for all the manufacturing operations. Every step of the manufacturing process has its own view of the product data. These dissimilar views make it difficult to return to the designer evaluative or corrective knowledge about the different processes.

As we move into the twenty-first century, new manufacturing technologies are needed to improve productivity and competitiveness. In our information and computer age, companies exchange and share information across the country. This capability is needed for manufacturing today's complex products such as automobiles, airplanes, ships, and buildings.

Multi-enterprise concurrent engineering will require the ability to store and retrieve product data far beyond the capability of the mechanical drawing. The replacement for the mechanical drawing that will allow revolutionary new engineering technologies is product data sharing. This new capability will make available to the designer knowledge about all other processes. It will process product data through automated computer-based techniques that allow for shared access among the life cycle processes in support of concurrent engineering. It will make available an integrated product data model that allows access to multiple views of the product.

STEP, as well as other new product data, data exchange and interface standards and their supporting technologies, must be implemented for this new product data sharing capability to be successful. *That is why concurrent engineering is impossible without standards.*

The critical concept is this: standards for product data and data exchange are important because they enable and facilitate an automated form of concurrent engineering that can be implemented in a computerized environment. This automated, or computer-aided, concurrent engineering provides a mechanism for multi-enterprise integration. As a result, the automated practice of concurrent engineering among manufacturing enterprises, their customers and their suppliers, including suppliers of technology as well as materials and components, would create a new kind of multi-enterprise concurrent engineering. This kind of multi-enterprise concurrent engineering could be achieved without the surrendering of historical forms of personal interactions currently practiced by workers and managers.

Accordingly, an automated approach to multi-enterprise concurrent engineering could be especially valuable in the commercial environment of the U.S. It could merge the many dynamic and innovative, mostly small, entrepreneurial companies along with larger manufacturing enterprises into an integrated and cooperative group

Yet although they would be integrated in the way they contribute their talents to the life cycles of products, participants in such diverse groups could remain individualistic and independent in the way they operate and manage their businesses. Although participating companies would work in an integrated fashion, and enjoy the benefits of concurrent engineering, their ability to retain their individual freedoms would preserve for them the benefits associated with the traditional strengths of U.S. commercial and individual diversity.

In these ways, multi-enterprise concurrent engineering could match the historically successful style of entrepreneurial innovation in the U.S. with the competitive and economic demands of today's global economy. The result could be the reemergence of U.S. manufacturing in world markets.

ACKNOWLEDGEMENTS

A substantial number of the concepts and results discussed in this chapter represent the accomplishments of staff members of the Factory Automation Systems Division in the Manufacturing Engineering Laboratory, and of other divisions at NIST. We are grateful for their technical contributions, as well as their leadership efforts in ensuring that manufacturing data interface standards will indeed become a reality. Many members of our Division provided us with information about their work for this chapter. We are grateful for their assistance. We especially thank, for their more direct contributions to our manuscript, Charles R. McLean, Manager of the CALS/PDES Project, which includes the NIST National PDES Testbed; Peter Brown, Manager of the Engineering Design Laboratory; and Cita Furlani, Leader of the Product Data Engineering Group, all in our Division. Also, we thank Sharon J. Kemmerer, who is in the Information Systems Engineering Division of the NIST Computer Systems Laboratory, for her many valuable suggestions that improved our manuscript.

REFERENCES

1. R. I. Winner, J. P. Pennell, H. E. Bertrand, and M. M. G. Slusarczuk, "The Role of Concurrent Engineering in Weapons System Acquisition," Institute for Defense Analyses, Report R-338, 175 pages, (December 1988).

2. "America's Choice: High Skills or Low Wages," Commission on Skills of the American Workforce, I. C. Magaziner, Chair (1990). (Available from National Center on Education and the Economy, 39 State Street, Suite 500, Rochester, NY 14614.)

3. A. Lowenstein and S. Schlosser, "DOD/Industry Guide to Automation of a Concurrent Engineering Process," Prospective Computer Analysts, Inc. Report CE-Guide, 114 pages (December 1988). (Available from Prospective Computer Analysts, Inc., 1800 Northern Blvd., Roslyn, NY 11576.)

4. J. L. Nevins and D. E. Whitney, eds., "Concurrent Design of Products and Processes: A Strategy for the Next Generation In Manufacturing," McGraw-Hill, New York, 1989.

5. "Application of Concurrent Engineering to Mechanical Systems Design," CALS Industry Steering Group, Technical Report 002, 97 pages (June 16, 1989). (Available from the National Security Industrial Association, Attn: CALS Industry Steering Group, Suite 300, 1025 Connecticut Ave. NW, Washington, DC 20036.)

6. C. W. Kelly, and J. L. Nevins, "Findings of the U.S. Department of Defense Technology Assessment Team on Japanese Manufacturing Technology," The Charles Stark Draper Laboratory, Report R-2161, for the Defense Advanced Research Agency/Information Science and Technology Office, Procurement Number MDA972-C-0027, 209 pages (June 1989).

7. H. M. Bloom, "The Role of the National Institute of Standards and Technology As It Relates To Product Data Driven Engineering," National Institute of Standards and Technology, Interagency Report 89-4078, 35

pages (April 1989). (Available from the National Technical Information Service (NTIS), Springfield, VA 22161.)

8. "A Framework for Concurrent Engineering," CALS Industry Steering Group (to be published). (Available from the National Security Industrial Association, Attn: CALS Industry Steering Group, Suite 300, 1025 Connecticut Ave. NW, Washington, DC 20036.)

9. J. Naisbitt, "Megatrends: Ten New Directions Transforming Our Lives," Warner Books, New York (1982).

10. "Emerging Technologies: A survey of Technical And Economic Opportunities," Technology Administration, U.S. Department of Commerce, 55 pages (Spring 1990).

11. "The Competitive Status of the U.S. Electronics Sector from Materials to Systems," International Trade Administration, U.S. Department of Commerce, 221 pages (April 1990). (Available from the Superintendent of Documents, U.S. Government Printing Office, Washington, DC 20402-9325.)

12. "Making Things Better: Competing in Manufacturing," Office of Technology Assessment, Congress of the United States, OTA-ITE-443, 241 pages (February 1990). (Available from the Superintendent of Documents, U.S. Government Printing Office, Washington, DC 20402-9325.)

13. K. B. Clark, W. B. Chew, and T. Fujimoto, "Product Development in the World Auto Industry: Strategy, Organization, Performance," Presentation to the Brookings Institution Microeconomics Conference, December 3, 1987. (Available from the Graduate School of Business Administration, Harvard University.)

14. C. H. Ferguson, "Computers and the Coming of the U.S. Keiretsu," *Harvard Business Review* **68**, No. 4, 55-70, (July-August 1990).

15. D. L. Wince-Smith, *in* "Debate: Can a Keiretsu Work in America?," *Harvard Business Review* **68**, No. 5, 197 (September-October 1990).

16. W. M. Henghold, G. C. Shumaker, and L. Baker, Jr., "Considerations for the Development and Implementation of PDES Within a Government Environment," Manufacturing Technology Directorate, Air Force Wright Aeronautical Laboratories, Report AFWAL-TR-89-8009, 145 pages (February 1989).

17. "Computer-aided Acquisition and Logistic Support (CALS) Program Implementation Guide," Department of Defense, Military Handbook MIL-HDBK-59A, September 1990.

18. K. A. Reed, D. Harrod, and W. Conroy, "The Initial Graphics Exchange Specification, Version 5.0," National Institute of Standards and Technology, Interagency Report 4412 (September 1990). (Available from the National Technical Information Service (NTIS), Springfield, VA 22161.)

19. "IGES/PDES Organization: Reference Manual," National Computer Graphics Association, Fairfax, VA, July 1990.

20. "STEP Part 1: Overview and Fundamental Principles," ISO TC184/SC4/WG1, Document N494 Version 3 (May 1990).

21. C. Furlani, J. Wellington, and S. Kemmerer, "Status of PDES-Related Activities (Standards and Testing)," National Institute of Standards and Technology, Interagency Report 4432, 18 pages (October 1990). (Available from the National Technical Information Service (NTIS), Springfield, VA 22161.)

22. "A Strategy for Implementing a PDES/STEP Data Sharing Environment," PDES, Inc. (to be published). (Available from the South Carolina Research Authority, Trident Research Center, 5300 International Blvd., N. Charleston, SC 29418.)

23. "EXPRESS Language Reference Manual," ISO TC184/SC4/WG1, Document N496 (May 1990).

24. "The STEP File Structure," ISO TC184/SC4/WG1, Document N279 (September, 1988).

25. "Common Data Model Subsystem, Data Model Subsystem, Vol. 5, Part 4, Information Modeling Manual IDEF1-Extended (IDEF1X)," Manufacturing Technology Directorate, Air Force Wright Aeronautical Laboratories, Report AFWAL-TR-86-4006 (1986).

26. S. N. Clark, "An Introduction to the NIST PDES Toolkit," National Institute of Standards and Technology, Interagency Report 4336, 8 pages (May 1990). (Available from the National Technical Information Service (NTIS), Springfield, VA 22161.)

27. M. Palmer, and M. Gilbert, "Guidelines for the Development and Approval of Application Protocols," Working Draft, Version 0.7, ISO TC184/SC4/WG4 Document N1 (February 1991).

28. M. Mitchell, "Testbed Development Plan: Validation and Testing System Development," National Institute of Standards and Technology, Interagency Report 4417, 38 pages (September 1990). (Available from the National Technical Information Service (NTIS), Springfield, VA 22161.)

29. "Bylaws of the Digital Representation of Product Data Standards Harmonization Organization," ANSI (to be published). (Available from the American National Standards Institute, 1430 Broadway, New York, NY 10018.)

30. M. Mitchell, Y. Yang, S. Ryan, and B. Martin, "Data Model Development and Validation for Product Data Exchange," National Institute of Standards and Technology, Interagency Report 90-4241, 13 pages (January 1990). (Available from the National Technical Information Service (NTIS), Springfield, VA 22161.)

31. M. J. McLay and K. C. Morris, "The NIST STEP Class Library (STEP into the Future)," National Institute of Standards and Technology, Interagency Report 4411, 20 pages (August 1990). (Available from the National Technical Information Service (NTIS), Springfield, VA 22161.)

32. C. Stark, and M. Mitchell, "Development Plan: Application Protocols for Mechanical Parts Production," National Institute of Standards and Technology, Interagency Report (to be published). (Available from the National Technical Information Service (NTIS), Springfield, VA 22161.)

33. W. F. Danner, "A Proposed Integration Framework for STEP (Standard for the Exchange of Product Model Data)," National Institute of Standards and Technology, Interagency Report 90-4295 (April 1990). (Available from the National Technical Information Service (NTIS), Springfield, VA 22161.)

34. W. F. Danner, and Y. Yang, "Generic Product Data Model (GPDM)," National Institute of Standards and Technology, Interagency Report, (to be published); also, "Generic Enterprise Data Model (GEDM)," National Institute of Standards and Technology, Interagency Report, (to be published). (Available from the National Technical Information Service (NTIS), Springfield, VA 22161.)

35. C. R. McLean, "Interface Concepts for Plug-Compatible Production Management Systems," Proceedings of the IFIP W.G. 5.7: Information Flow in Automated Manufacturing Systems, Gaithersburg, MD, August 1987; reprinted in *Computers in Industry* **9**, 307-318 (1987).

36. H. M. Bloom, C. Furlani, M. Mitchell, J. Tyler, and D. Jefferson, "Information Resource Dictionary System: An Integration Mechanism for Product Data Exchange Specification," National Institute of Standards and Technology, Interagency Report 88-3862, 15 pages (October 1988). (Available from the National Technical Information Service (NTIS), Springfield, VA 22161.)

37. T. R. Kramer, "Enhancements to the VWS2 Data Preparation Software," National Institute of Standards and Technology, Interagency Report 89-4201, 58 pages (November 1989). (Available from the National Technical Information Service (NTIS), Springfield, VA 22161.)

38. H. T. Moncarz, "Architecture and Principles of the Inspection Workstation," National Institute of Standards and Technology, Interagency Report 88-3802, 46 pages (September 1990). (Available from the National Technical Information Service (NTIS), Springfield, VA 22161.)

39. J. S. Tu and T. H. Hopp, "Part Geometry Data in the AMRF," National Institute of Standards and Technology, Interagency Report 87-3551, 16 pages (April 1987). (Available from the National Technical Information Service (NTIS), Springfield, VA 22161.)

40. D. Libes and E. J. Barkmeyer, "The Integrated Manufacturing Data Administration System (IMDAS)--An Overview," *International Journal of Computer Integrated Manufacturing* **1**, 44-49 (1988).

41. S. Rybczynski, E. J. Barkmeyer, E. K. Wallace, M. L. Strawbridge, D. E. Libes, and C. V. Young, "AMRF Network Communications," National Institute of Standards and Technology, Interagency Report 88-3816, 206 pages (June 1988). (Available from the National Technical Information Service (NTIS), Springfield, VA 22161.)

42. A. B. Feeney, "Engineering Design Laboratory Guide," National Institute of Standards and Technology, Interagency Report 4519, 13 pages (February 1991). (Available from the National Technical Information Service (NTIS), Springfield, VA 22161.)

43. Y. T. Lee, "On Extending the Standard for the Exchange of Product Data to Represent Two-Dimensional Apparel Pattern Pieces," National Institute of Standards and Technology, Interagency Report 4358, 27 pages

(June 1990). (Available from the National Technical Information Service (NTIS), Springfield, VA 22161.)

44. C. R. McLean, "National PDES Testbed Strategic Plan 1990," National Institute of Standards and Technology, Interagency Report 4438, 73 pages (October 1990). (Available from the National Technical Information Service (NTIS), Springfield, VA 22161.)

45. S. J. Kemmerer, "Development Plan: Conformance Testing System," National Institute of Standards and Technology, Interagency Report (to be published). (Available from the National Technical Information Service (NTIS), Springfield, VA 22161.)

46. J. E. Fowler, "STEP Production Cell Technical Development Plan," National Institute of Standards and Technology, Interagency Report 4421, 29 pages (September 1990). (Available from the National Technical Information Service (NTIS), Springfield, VA 22161.)

47. S. P. Frechette and K. Jurrens, "Development Plan: Product Data Exchange Network," National Institute of Standards and Technology, Interagency Report 4431, 27 pages (September 1990). (Available from the National Technical Information Service (NTIS), Springfield, VA 22161.)

48. S. Ressler and S. Katz, "Development Plan: Configuration Management Systems and Services," National Institute of Standards and Technology, Interagency Report 4413, 28 pages (September 1990). (Available from the National Technical Information Service (NTIS), Springfield, VA 22161.)

49. A. Jones, E. Barkmeyer, and W. Davis, "Issues in the Design and Implementation of a System Architecture for Computer Integrated Manufacturing," *Int. J. Computer Integrated Manufacturing* **2**, No. 2., 65-76 (1989).

50. A. Jones, ed., "Proceedings of CIMCON '90," National Institute of Standards and Technology, Special Publication 785, 528 pages (May 1990). (Available from the National Technical Information Service (NTIS), Springfield, VA 22161.)

51. A.-W. Scheer, "Enterprise-Wide Data Modelling," Springer-Verlag, Berlin 1989.

52. "Open System Architecture for CIM," Springer-Verlag, Berlin 1989. (Available from ESPRIT Consortium AMICE, 489 Avenue Louise, Bte. 14, B-1050 Brussels, Belgium.)

53. "US TAG to ISO TC 184: Industrial Automation Systems," US TAG to ISO TC 184, Document N 244 (August, 1989).

54. "Improving Engineering Design: Designing for Competitive Advantage," Committee on Engineering Design Theory and Methodology, Manufacturing Studies Board, Commission on Engineering and Technical Systems, National Research Council (1991). (Available from National Academy Press, 2101 Constitution Ave., Washington, DC 20418.)

55. "Report of the National Critical Technologies Panel," National Critical Technologies Panel, W. D. Phillips, Office of Science and Technology Policy, Chair (1991). (Available from The National Critical Technologies Panel, 1101 Wilson Boulevard, Suite 1500, Arlington, VA 22209 and the National Technical Information Service (NTIS), Springfield, VA 22161.)

NEW CONCEPTS OF PRECISION DIMENSIONAL MEASUREMENT FOR MODERN MANUFACTURING

DENNIS A. SWYT

Precision Engineering Division
National Institute of Standards and Technology
Gaithersburg, Maryland 20899

I. INTRODUCTION

A hallmark of modern products is the high precision of the dimensions of their functional parts — high, that is, compared to that of their less-modern contemporaries. Since the Industrial Revolution, such has been the case. Modern products of the last century included certain factory-produced small arms, the precision of whose parts allowed them to be interchangeably assembled in the then-new system of mass production. Modern products of this century are video cassette recorders, the precision of whose parts allows read-write heads to be aerodynamically flown over the recording medium at microscopic altitudes.

Since it is modern products which sell best, for U.S. manufacturing firms seeking to compete better against evermore sophisticated foreign producers, the ability to realize precision higher than that of their competitors is essential. Precision, however, is relative not only to the capabilities of the competition, but also to the size of the functional parts in question. Comparable degrees of high precision are a common end for a wide range of manufacturers, from producers of the very large, such as commercial aircraft, to producers of the very small, such as nanoelectronic devices.

This chapter presents a new scheme for the analysis of the dimensional capabilities of measuring machines and machine tools aimed at the realization

CONTROL AND DYNAMIC SYSTEMS, VOL. 45

of higher degrees of precision in the manufacture of dimensioned parts. Successive sections deal with the range and types of dimensional measurements, their relation to the basic unit of measure, the means for making such measurements, and the errors associated with them. Application of these principles is illustrated by a detailed analysis of the use of a laser-based coordinate measuring machine for the measurement of a physical part. Also included is an indication of how the identical analysis may be applied to a machine tool which shapes such parts.

II. THE SCOPE OF DIMENSIONAL MEASUREMENTS

Manufacture of today's most modern products requires the ability to carry out precise dimensional measurements over a wide scope, one which includes subtly different types and astonishingly different scales. Dimensional measurements of all types and scales, however, are expressed in terms of the basic unit of length.

A. The Range of Dimensional Measurements

The range over which precision dimensional measurements for the character-ization of manufactured goods are made today spans nearly twelve orders of magnitude, segmented into relatively well recognized regimes:

1. The Macro-Scale

On the large-size end of the range are products which are not so much manufactured as constructed, that is, large-scale products often assembled on site rather than manufactured on line. Macro-scale products such as ships, aircraft, and spacecraft can have longest dimensions of the order of a hundred meters.

2. The Mid-Scale

In the mid-size of the range are products the parts of which are manufactured and assembled in the familiar factory environment, including a host of commonplace products from automobiles to machine screws. Mid-scale products have characteristic dimensions of the order of multiple- to fractional-meters.

3. The Micro-Scale

On the conventional small-size end of the range are products which are manufactured within special machines such as optical, electron-beam, and x-ray lithography systems. Less familiar products such as ultra-large-scale-integration

(ULSI) microelectronic devices. Micro-scale products have characteristic dimensions often of less than a micrometer.

4. The Nano-Scale

Just emerging now is technology for the manufacture of products of the "nano-scale," including nanoelectronic (as opposed to microelectronic) transistor-type devices 100 times smaller than those in present commercial production and mechanical devices such as ultra-miniature pumps capable of implantation into blood vessels and automatically metering out drugs. Nano-scale structures and devices have features composed of small numbers of atoms and molecules and have characteristic dimensions of the order of nanometers (10^{-9}m or 40 billionths of an inch). Achievement of high precision in dimensional measurements at the nanometer scale demands utmost attention to physical-scientific, engineering and metrological principles [1].

B. The Types of Dimensional Measurements

In the characterization of today's most modern manufactured products, precision dimensional measurements are carried out not only over a wide range of dimensions but over a range of types, each of which, while intimately related to the others, has a uniquely distinct role in measurement.

1. Position

a. *Geometrical Concept of Position*: In geometry, position is the location of a point relative to either: another point, such as the origin of a one-dimensional coordinate system; a line, such as an axis of a two-dimensional coordinate system; or a plane, such as the reference surface in a three-dimensional coordinate system. For a three-dimensional cartesian system, the formally explicit expression for geometric position is a vector quantity:

$$p = p \ (xx, yy, zz) \tag{1}$$

where x, y, and z represent the coordinates along unit vectors x, y, z relative to the (suppressed null-vector) origin of the coordinate system.

b. *Physical Concept of Position*: In physics, position is the location in space of a physical body. In the rigid-body approximation, where the configuration of that body is fully specified by six generalized coordinates (three linear and three rotational) corresponding to its six degrees of freedom, position is given by the three linear coordinates. Note that since both a rigid body and a reference frame have the same six degrees of freedom, locating a body relative to a reference frame is fully equivalent to locating one rigid body relative to another. In a three-dimensional cartesian system, the formally explicit expression for physical position is a vector quantity:

$$P = P\ (XX,\ YY,\ ZZ) \tag{2}$$

where X, Y, and Z represent the coordinates of the location of one physical body relative to the location of another physical body which acts as the reference frame of the coordinate system.

c. *Measurement Concept of Position*: In measurements, position is the assigned numerical value of the length of path between the single point which describes the location of a physical body and the origin, axis, or plane of the one-, two-, or three-dimensional coordinate system expressed in terms of the standard unit of measure. In a three-dimensional cartesian system, the formally explicit expression for measurement position is a vector quantity:

$$P = P\ (\ XX,\ YY,\ ZZ) \tag{3}$$

where X, Y, and Z, expressed in meters, represent the coordinates of the location of one physical body relative to the location of another physical body which acts as the reference frame of the coordinate system. Measurements of position are made, for example, in precision surveying of geodetic coordinates relative to control points and in measurements of the locations of points on objects by means of coordinate measuring machines.

2. Displacement

a. *Geometrical Concept of Displacement*: In geometry, there is nothing fully equivalent to displacement. However, the physical notion of displacement does involve the geometrical notion of translation, where translation is the transformation of a point at one position into a point at a different position by means of the operation:

$$t = p_f - p_o \tag{4}$$

where t is the vector of translation and p_o and p_f are the initial and final position vectors of the point.

b. *Physical Concept of Displacement*: In physics, displacement is the change in location in space of a physical body with time, that is, its movement from one location to another, and is described by:

$$T = P_f\ (t_1)\ -\ P_o\ (t_2), \tag{5}$$

where T is the translation representing the change in position with time of a single body from position $P_o(t_1)$ at time one to position $P_f(t_2)$ at time two.

c. *Measurement Concept of Displacement*: In measurements, displacement is the assigned numerical value of the length of path between the location of a physical body at an initial time and its location at a final time expressed in terms of the standard unit of measure. In a three-dimensional cartesian system, the formally explicit expression for measured displacement **D**, expressed in meters, would be a vector quantity:

$$\mathbf{D} = \mathbf{P_f}(t_1) - \mathbf{P_o}(t_2), \tag{6}$$

where the difference between $P_o(t_1)$ and $P_f(t_2)$ represents the change in location, between times one and two, of one physical body relative to the location of another physical body (which acts as the reference frame of the coordinate system). Measurements of displacement are made, for example, in the characterization of the travel of a stage by means of a laser interferometer, where the stationary parts of the interferometer comprise the body which defines the reference frame.

3. Distance

a. *Geometrical Concept of Distance*: In geometry, distance is the separation between two points. In a three-dimensional cartesian system, the formally explicit expression for geometric distance would be a scalar quantity:

$$d = |\ p_1\ (x_1x,\ y_1y,\ z_1z) - p_2\ (x_2x,\ y_2y,\ z_2z)\ |, \tag{7}$$

where x, y, and z represent the values of measured coordinates relative to the usually-suppressed null-vector origin of the coordinate system.

b. *Physical Concept of Distance*: In physics, distance is the separation in space of the locations of two objects. In a three-dimensional cartesian system, the formally explicit expression for physical distance would be a scalar quantity:

$$d = |\ P_1(X_1X,\ Y_1Y,\ Z_1Z) - P_2(X_2X,\ Y_2Y,\ Z_2Z)\ |, \tag{8}$$

where d is the distance between two bodies located at positions P_1 and P_2 respectively.

c. *Measurement Concept of Distance*: In measurements, distance is the assigned numerical value of the length of path between the location of one physical body and the location of another, expressed in the standard unit of measure.

A formally explicit expression for measured distance, expressed in meters, would be a scalar quantity:

$$d = |\mathbf{P}_1(t_1) - \mathbf{P}_2(t_1)|, \tag{9}$$

where $\mathbf{P}_1(t_1)$ and $\mathbf{P}_2(t_1)$ represent the locations of two different bodies at the same time. Measurements of distance are made, for example, in the center-to-center separation of engine cylinders and microelectronic circuit elements and in the calibration of the spacings of the successive graduations of various types of what are usually called "length scales," ranging from hundred-meter-long survey tapes to micrometer-long "pitch" standards for optical- and electron-microscope measurement systems.

4. Extension

a. *Geometrical Concept of Extension:* In geometry, extension is the line segment between two points which lie on a surface which is topologically equivalent to a sphere such that all points on the line segment near the surface are either interior to or exterior to the surface. Equivalently, extension is the line segment connecting two points on a surface which have exterior normals with components that are anti-parallel. A formally explicit expression for physical extension e is:

$$e = |r_{b1}(\psi) - r_{b2}(\psi + 180°)|, \tag{10}$$

where $r_{b1}(\psi)$ and $r_{b2}(\psi + 180°)$ represent the locations of points on a surface having exterior normals 180° apart.

b. *Physical Concept of Extension:* In physics, extension is the amount of physical space that is occupied by, or enclosed by, a material object. A formally explicit expression for physical extension is:

$$E = |R_{b1}(\psi) - R_{b2}(\psi + 180°)|, \tag{11}$$

where $R_{b1}(\psi)$ and $R_{b2}(\psi + 180°)$ represent respectively the locations of the opposite-facing boundaries of the physical object which, for convenience, may be expressed in terms of a vector radius from the centroid of the body.

c. *Measurement Concept of Extension:* In measurements, extension is the assigned numerical value of the length of path between the location of one boundary of a material object relative to the location of another opposite-facing boundary, expressed in terms of the standard unit of measure. A formal expression for measured extension is:

$$e = |\ P_{b1}(\psi) - P_{b2}(\psi + 180°)\ | \tag{12}$$

where e is the extension and P_{b1} and P_{b2} represent the locations of opposite-facing boundaries of an object. Extension-type dimensions of objects include height, width, diameter, and also "length". Measurements of extension are made in general to achieve physical fit between mating parts. Prototypical examples of extension measurements are those in the automotive industry wherein the outside diameter of piston rings and the inside diameter cylinder walls are measured to insure slip fit. For analogous reasons, extension measurements are also made in the microelectronics industry on critical elements of microcircuit devices. Table 1 below summarizes key aspects of the various types of dimensional measurements just described.

Table 1.   **Description of the Quantity and Nature of the Length of Path To Be Measured for Each of the Four Dimensional Measurement Types**

Dimension	Quantity	The Length of Path Is That Between:
Position	Vector **P**	The location in space of a single object and an origin of coordinates (equivalent to a second, reference object)
Displacement	Vector **D**	The change in location of one object from time t_1 to time t_2
Distance	Scalar d	The locations of two objects
Extension	Scalar e	The locations of opposite-facing boundaries of an object

In familiar terms, each type of "length" relates to a different, but interrelated, concern. For example, the "length" of an automobile (i.e. its extension) describe how big a parking space is needed, the "length" of a racetrack (i.e. its distance) describes how far it is between start and finish lines, and the "length" of a stage's travel (i.e. its displacement) describes how far its carriage can move.

III. THE BASIS OF DIMENSIONAL MEASUREMENTS

What "lengths" of all scales and types have in common is that each is measured and expressed in terms of a unit of length. For that unit to be the internationally standard unit of length, a formal definition must be operationally realized and transferred.

A. Realization of the Unit of Length

The function of a unit of length is to provide a metric for the physical space in which dimensional measurements are to be made. The function of a standard unit of length is to provide a metric which can be used in common in international science and trade.

1. The Formal Definition of the Meter

By treaty, the standard of length used in international science and trade is the meter. The meter is part of the International System of Units (SI) and is referred to as the fundamental, or SI, unit of length. Since 1983, the formal SI definition of the meter as the unit of length has been:

> "*The meter is the length of path travelled by light in vacuum in the time interval 1 / 299 792 458 of a second*" [2].

No longer based on lines scribed on a platinum-iridium "meter" bar (as it was from 1887 until 1960), the meter today is the distance which light travels in free space in (approximately) three billionths of a second.

2. The Derived Nature of the Meter

The new definition of the meter makes it in effect a derived quantity governed by the length-time relation:

$$L = c \cdot T, \tag{13}$$

where L is the defined unit of length, c is the defined constant speed of light, and T is the defined unit of time. The second is the SI unit of time and is defined in terms of a fixed number of counts of the frequency of oscillation of specific atomic states of a cesium-beam atomic clock.

3. The Operational Realization of the Meter

The meter is operationally realized through a chain of intercomparisons of the frequencies of various microwave-, infrared- and visible-wavelength lasers, based on the relationship of the frequency of electromagnetic waves and their wavelength:

$$f \cdot \lambda_o = c \, , \tag{14}$$

where f is frequency, λ_o is vacuum wavelength, and c is value of the speed of light (that is, $2.99792458 \cdot 10^8$ m/sec). Based on this relationship, the ultimate limit on the uncertainty associated with the realization of the meter is that of the cesium clock, which is $10^{-13.}$ At present, for dimensional measurements access in principle to the frequency cesium-beam atomic clock is through a chain of intercomparisons of wavelengths of microwave, infrared and visible sources by which the frequency of visible-wavelength lasers is determined.

B. The Transfer of the Metric to Physical Objects

In practical dimensional measurements, the means by which the SI unit of length as the metric of physical space is transferred to measurements of physical objects -- from automobile piston rings to microelectronic circuit elements -- is through a system which successively involves visible light of known wavelength, displacement interferometry using that light, coordinate measurement systems based on that interferometry, and measurements on physical objects using such coordinate measurement systems.

1. Visible Light of Known Wavelength

For dimensional measurements, the practical primary standard of length today is the vacuum wavelength of the 633-nm red-orange line of an iodine-absorption-stabilized helium-neon laser. One may realize the SI-unit-based wavelength, to a particular accuracy, by buying a commercial helium-neon laser stabilized to a corresponding degree.

2. Displacement Interferometry

The means by which propagating laser light, which defines the metric of physical space, is coupled to dimensional measurements in that space is through interferometric measurements of displacement.

The prototypical displacement interferometer is the two-beam, Michelson interferometer. The Michelson interferometer consists of a laser light source, beam splitter, fixed reflector (either a plane mirror or retroreflector), moving reflector and detector. In this type of interferometer the translation of the moveable reflector, which must be parallel to the axis of propagation of the laser light, is measured in terms of whole and fractional numbers of fringes generated by the interference of the beams in the stationary and moving arms of the interferometer.

The typical displacement interferometer of today employs the heterodyne principal for measuring the change in optical phase between the two beams by having one frequency-shifted with respect to the other. A widely-used commercial heterodyne interferometer employs a two-frequency laser and the displacement of a plane mirror or retroreflector is measured by means of two channels of detectors, doublers, counters, subtractors and a calculator [3].

By whatever means the phase difference between the reference and moving-mirror beams is measured, the operational link between the SI-unit-carrying wavelength of visible light and a displacement measurement is through the interferometer equation:

$$D = L_1 - L_2 = N\lambda_o / 2n_m \cdot \cos\theta, \qquad (15)$$

where D is the displacement, $L_1 - L_2$ is the change in length (i.e., extension) of the interferometer cavity, λ_o is the vacuum wavelength of the laser light used, N is the real number describing the counted integer-order and measured fringe-fraction of the interference, n_m is the index of refraction of the medium, and θ is the angle between the optical axis of the interferometer and the direction of propagation of the incident light. In multi-pass fringe-fractioning interferometers:

$$D = M\lambda_o / 2nmK \cdot \cos\theta, \qquad (16)$$

where M is the counted integer-order and measured fringe-fraction of the interference, m is the number of passes of the light through the interferometer, and K is the number of electronic sub-divisions of the interference fringes (typically a power of two ranging from 16 to 1024 depending on the system).

For the high-resolution commercial heterodyne interferometer system described above, the limit of resolution, that is, the least count in displacement measurement is about λ /400 or 1.5 nm. Under development now at NIST is a specially optimized heterodyne interferometer intended to realize a least count of twenty times higher, that is, λ /8000 or 0.07 nm.

• • •

Note that, as described here, in a laser displacement-interferometry system, it is the value of the vacuum wavelength of the laser light which provides the metric of physical space as the fundamental unit of length. It is the overall interferometer system doing the subdivision of fringes which provides the "scale," that is, the metric-less subdivisions of that unit. In common usage, however, displacement-measuring devices, such as interferometer systems, are often called "scales" and perform both of the functions described here: subdividing the axes (that is, providing the scale for an axis) and carrying the metric (that is, providing the standard unit of length, relative to which the "scales" have somehow been appropriately calibrated).

3. The Coordinate Measurement System

After the wavelength of laser light and displacement interferometry, the next link to dimensional measurements of physical objects is a coordinate measurement system. A coordinate measurement system is necessary for the measurement of "position," that is, the location of a single point in space.

To establish a coordinate system, one must establish the means for realizing the geometry for physical space, including: *axes* (one, two, or three depending on the dimensionality of the system desired; *an origin* (the null-vector point from which radiate the axes of the coordinate system); *a scale along each of the axes* (that is, the graduation of each axis into subdivisions); and *a metric* (that is, a single measure of distance in the space of the reference frame to which the graduations of the axes relate).

In modern, automation-oriented manufacturing, the means by which the geometry of the coordinate measurement system is realized and precision dimensional measurements are made, is the coordinate measuring machine (CMM). A coordinate measuring machine embodies a coordinate system by means of its essential functional components, which are its frame, scales, carriages and probes as outlined in Table 2.

Today's high-performance three-dimensional coordinate measuring machines take on a wide variety of forms depending on the specific way in which each of these functional components is embodied. Frames can be of the bridge or cantilever-arm type. Scales can be ruled-glass optical encoders or enclosed-path laser interferometers. Carriages can be translating and rotating tables and moving bridges. Probes can be touch-fire or analog mechanical-contact types or capacitive, optical, or machine-vision non-mechanical-contact types.

Whatever particular form a coordinate measuring machine takes on, its fundamental function is to provide measurements of locations in space, that is, of position. However, just as a manufacturer wants to buy holes not drills, users of CMMs in manufacturing want measurements of dimensions of objects, such as those of their products, not coordinates of locations in space.

4. The Physical Object To Be Measured

The terminus of the series of linkages, which allow dimensional measurements on physical objects in terms of the SI unit of length, is the object itself. Now there are two different kinds of features on objects corresponding respectively to the distance-type and the extension-type of dimensional measurements described above.

a. *Distance-Type Features of Objects*: Distance-type features, include, for example, the center-to-center spacings of cylinders of an automobile or spacing of left-edges of successive microcircuit elements. While, in fact, always to some

Table 2. **The Principal Functional Components of a Coordinate Measuring Machine (as well as of a Machine Tool)**

Frame	The frame is the geometry-generating structure, which is the means for embodying the origin, axes, and angular relationships which comprise a coordinate system.
Scales	The scales are the displacement-measuring devices, which are the means for realizing the metric-based graduations along each axis of the frame.
Probe	The probe is the sensor system, which is the means for linking the boundary of the object to be measured to the frame and scales; in a machine tool, the "probe" is a material-moving element rather than a material-locating element as it is in a measuring machine.
Carriage	The carriage is the complete motion-generating system, which is the means for translating the object to be measured relative to the probe, frame and scales.

degree extended, distance-type features are treated as extensionless points. Thus, the distance between automobile cylinders is measured in terms of their centroids. Similarly, the distance between the moon and the earth is center to center. Measurements of distance type features are inherently, in effect, point to point.

b. *Extension-Type Features of Objects:* Extension-type features include, for example, the diameters of the automobile cylinders or the widths of the microcircuit elements. This type of feature must be treated as made up of boundaries that must be approached from opposite directions as in calipering, in which one locates two points, one on each and measures the "distance" between them. However, it is in the essential nature of extension measurements that these boundaries are opposite facing. As will be seen below, it is the practical and theoretical difficulty in locating properly such boundaries that makes extension the most difficult type of dimensional measurement.

• • •

In sum, successful measurement of, for example, the diameter of an automobile cylinder or the width of a microelectronic element in terms of the SI unit of length requires dimensional measurements of each type in succession: displacement, position, distance, and, finally, extension. The measure of success in this succession of measurements at each level and in total is accuracy achieved.

D. Standards-Imposed Limits to Accuracy

The accuracy in practical dimensional measurements depends on the availability of standards for the calibration of the measuring machines by which measurements are made. Attainable accuracy is ultimately limited by national standards laboratories in measurements involving a realization of the meter, displacement interferometry, reference coordinate measuring machines, and calibrated artifacts.

1. The Limit of Realization of the Meter

The ultimate limit to accuracy in practical dimensional measurements is the accuracy with which national standards laboratories can realize the definition of the meter. At present that limit is the uncertainty with which the visible wavelength of the iodine-stabilized HeNe laser is known, that is, 10^{-9} compared to the 10^{-13} of the atomic clock. The meter as the unit-of-length metric of that space can be accessed directly then through the vacuum wavelength of the HeNe laser. Table 3 below shows the effective wavelength and uncertainty associated with HeNe lasers stabilized by various means.

As indicated in Table 3, the fractional uncertainties ($\Delta\lambda/\lambda$) associated with commercially available HeNe lasers range from the 10^{-6} of a free-running device, through the 10^{-7} specified stability of certain commercial lasers, and 10^{-8} short-term stability of some such devices, up to the 10^{-9} accuracy of the visible wavelength of the iodine-stabilized laser as defined in the documents associated with the definition of the meter and realized in laboratory devices.

2. The Limit of Displacement Interferometry

After the ultimate limit to accuracy of dimensional measurements imposed by the realization of the meter in terms of the wavelength of visible light, the next limit to that accuracy is the degree to which measurements of displacement can be made by means of optical-wavelength interferometry.

Based on work by national standards laboratories, the limit to the accuracy of interferometric displacement measurements at a displacement of a meter is estimated to be: 10^{-4} in air uncompensated for pressure, temperature and humidity; 10^{-7} in fully compensated ambient air; and 10^{-8} in standard dry air.

Table 3.   Comparison of the Value and Associated Fractional Uncertainty of the Vacuum Wavelength of Helium-Neon Lasers of Various Degrees of Stabilization

Wavelength Value	Uncertainty	Type of Stabilization
632. 991 nm	10^{-6}	Free-running
632. 9914	10^{-7}	Any one of a number of
632. 99139	10^{-8}	opto-electronic techniques
632. 991398	10^{-9}	Iodine Absorption Cell

Based on results by a variety of workers, a reasonable estimate for the practical limit for displacement interferometry in vacuum is 10^{-10}.

3. The Limit of Coordinate Measurement

The next factor to limit the accuracy of dimensional measurements after the realization of the meter and displacement interferometry is the ability to embody a coordinate system in a measuring machine. At present, the highest-accuracy coordinate measurement is achieved by national standards laboratories on special-purpose one- and two-axis measuring machines.

At the U. S. National Institute of Standards and Technology, a machine for the calibration of line scales over the range 1 μm to 1 m has associated with it a total uncertainty of measurement U_T given by:

$$U_T = 10 \text{ nm} + 10^{-7} \cdot s_x , \qquad (17)$$

where s_x is the separation of any two graduations on the scale. At one meter, the positional error of this machine is approximately $1 \cdot 10^{-7}$ [4]. For a two-axis machine used for the calibration of grid plates up to 600 mm square, the corresponding positional error in routine use is 0.5 μm, corresponding to about $5 \cdot 10^{-7}$ [5].

Currently unmet is the need in the microelectronic industry for an accuracy of 50nm accuracy over 250 mm, that is, $2 \cdot 10^{-7}$ [6]. Because of a trend to even more stringent requirements, for the regime of nanotechnology, there is under development at the NIST a "Molecular Measuring Machine"[7], which is a planar 50-by-50 mm xy-coordinate measuring machine (with a 100 μm z-axis) which has a design-goal positional capability of:

$$U_T = 1 \text{ nm} = 10^{-8} @ 70 \text{ mm}. \tag{18}$$

4. The Limit of Physical Artifact Standards

The limiting error in many practical dimensional measurements is the accuracy of artifact standards (that is, physical objects), dimensions of which have been calibrated by a national standards laboratory or by a secondary laboratory referenced to it. Accuracies representative of various types of artifact standards calibrated by NIST are shown in the Table 4 below.

Table 4.   The Type, Range, and Accuracies of Some Dimensional Standards Calibrated by the NIST

Dimensional Std	Type	Range	Accuracy
Survey Tapes	d	1 mm to 300 m	5.0 μm + $10^{-6} \cdot d$
Gauge Blocks	d	1 mm to 0.5 m	0.025 μm to 0.15 μm
Line Scales	d	1 μm to 1 m	0.01 μm + $10^{-7} \cdot d$
Photomask Lines	e	0.5 μm to 20 μm	0.05 μm
Polymer Spheres	e	0.3 μm to 3 μm	0.01 μm
Thin-Film Steps	d	0.02μm to 10 μm	0.003 μm to 0.2 μm

Note that because of their idiosyncratic nature and the intractability of dealing with the very object-specific boundary-location errors, NIST calibrates extension-type objects, such as integrated-circuit photomask linewidths [8], only when the national interest justifies the major long-term theoretical and experimental effort required.

Note also that while NIST provides measurement services for calibrations of distance and extension (which involve associating the unit of length with locations on physical objects), it does not calibrate displacement or position (which involve associating the unit of length with a device-based process [9].

IV. ASSESSING DIMENSIONAL MEASUREMENTS

As suggested earlier, manufacture of modern products demands dimensional measurements of not only exceptional precision, that is, the closeness together, but also of exceptional accuracy, that is, closeness to a true value [10].

Especially in high-technology products including, for example, microwave resonators for communication systems and x-ray optics for microelectronic lithography, accuracy is required because the dimension of device features is dictated not just by design convention but by the laws of physics. As a result, dimensions of features must not only be very close to *a* value but to *the* value. This section looks at the assessment of dimensional measurements in terms of statistical characterization of error, formal-theoretical characterization of error, compounding of error and standards-imposed limits of error.

A. Statistical Characterization of Error

Physical measurement is itself a process, like manufacturing, which produces output. The output of a measurement process is numbers, that is, the measurement results. The quality of these numbers can be assessed and characterized by means of measurement statistics in terms of precision and accuracy. The manufacture of modern products demands dimensional measurements that are of both high precision and high accuracy, however one chooses to describe that combined state.

1. Precision-vs-Accuracy

a. *Precision*: As a concept, precision conveys the notion of closeness together. As such, it is the principal figure of merit for the achievement of higher quality in the sense of lower variability. Precision is conventionally represented numerically in terms of the standard deviation from the mean for a number of measurements presumed to be randomly drawn from a large population describable by a Gaussian distribution function. The standard deviation for a single such measurement σ_i is given by:

$$\sigma_i = [\ \Sigma_i \ (\ell_i - \ell_m)^2/(n - 1) \ ]^{1/2}, \tag{19}$$

where $\ell_m = \Sigma_i \ell_i/n$ and the summations Σ_i are over the number of measurements n. Where σ_i is a measure of the precision of one measurement, σ_m is a measure of the precision of the mean of n measurements and is given by:

$$\sigma_m = \sigma_i / n^{1/2}. \tag{20}$$

Precision is given in terms of either one-, two- or three-sigma values for single measurements, corresponding respectively to confidences of 68.3%, 95.5%, and 99.7%. National standards laboratories typically use $3\sigma_m$ for characterizing dimensional standards which they calibrate. Instrument manufacturers often use $2\sigma_m$ for characterizing dimensional-measurement devices which they sell.

b. *Accuracy*: As a concept, accuracy conveys the notion of closeness to a true value. As such, accuracy is the principal figure of merit for the achievement of interchangeability of components and physical operation in high-technology devices. In those situations, actual not relative dimensions determine functional performance. Accuracy is conventionally represented numerically in terms of an estimate of the maximum that a measurement would likely be different from a true or standard value. In metrological terms, accuracy is the estimate of the degree to which measurements are free from systematic error and is represented by the term s.e.

2. Error or Total Uncertainty

The metrological term for a combined measure of precision and accuracy is uncertainty or error. Total uncertainty includes a fully-described estimate of systematic error and fully-described measure of precision. NIST, along with a number of the national standards laboratories of other countries, typically uses the three-sigma of the mean in the assignment of a total uncertainty given by:

$$U_T = \text{s.e.} + 3\sigma_m. \tag{21}$$

Obviously, informed judgment must be made in how to sum and combine systematic and random error contributions. For example, the magnitudes of systematic errors may be added algebraically, that is, as signed quantities, if the signs are definitely known or arithmetically, that is, without signs, if not. Random errors may be added in quadrature if known to be uncorrelated or arithmetically if not.

Throughout this chapter, the term "error" is used synonymously with uncertainty U_T as described above and is denoted by the Greek character delta, upper case (Δ) for total-type and lower (δ) for type-specific.

3. Additive-Multiplicative Representation

Dimensional-measurement errors from different sources can be either additive in nature, that is, appearing as incremental off-set, or they may be multiplicative, that is, appearing as a length proportional. Total error can be of the form:

$$\Delta \ell = A + B \cdot \ell, \tag{22}$$

where A is the additive error and B the multiplicative.

a. *Added and Multiplicative Components:* Given a system in which errors from a number of sources are to be combined to estimate the total error in a length-based dimensional measurement of a given type, then:

$$\Delta \ell = (\Sigma \delta \ell_{Ai}) + (\Sigma \delta \ell_{Bi}) \cdot \ell, \tag{23}$$

where A now is the sum Σ_i over individual additive contributions $\delta\ell_{Ai}$ and B the sum Σ_i over individual multiplicative contributions $\delta\ell_{Bi}$, where A_i and B_i can be comprised of either or both systematic errors and random errors.

b. *Manufacturers' Statements of Accuracy*:   Many   commercial producers of measuring machines conventionally quote an accuracy for the performance of their products in additive-multiplicative form, as in, for example:

$$\Delta L = 2\mu m + [\ L(mm)/500mm\ ]\ \mu m, \tag{24}$$

which means that over the 1m-travel of the machine, the accuracy of single measurements of position would range from a minimum of $2\mu m$ to a maximum of $4\mu m$.

B. Formal-Theoretical Characterization of Error

Simple assessments and statements of accuracy which put errors into the simple additive-multiplicative form lump two types of additive errors, constant off-sets and non-linearities, and provide an inadequate basis for identification of the sources of error. For diagnostic purposes, the better approach is to consider the most general case, that of a dimensional measuring system that produces output which includes linear and non-linear terms in the form:

$$\ell_{obs} = a + (1+b) \cdot \ell_t + c(\ell_t^n), \tag{25}$$

where ℓ_{obs} is the measured value, ℓ_t is the true length, a and b are constants and $c(\ell_t^n)$ is the sum of all non-linear terms. The error in measured values $\Delta\ell$ is then given by:

$$\Delta\ell = \ell_{obs} - \ell_t = a + b \cdot \ell_t + c(\ell_t^n). \tag{26}$$

Each of the error components a, b and c describes a specific type of error source and is a measure of the degree to which the measurement system has realized a fundamental requirement of measurement.

1. The Fundamental Axioms of Measurement

For dimensional measurements to yield physically meaningful results, they must conform to axiomatically fundamental requirements given in the *Philosophical Foundations of Physics* by Rudolf Carnap [11] and provided a more accessible account by Simpson [12]. These axiomatic requirements are that a measuring system must be able to rank order objects along the dimension of measurement and to reproduce properly the unit, the zero, and the scale of that dimension.

a. *The Rank-Order Operator:* The rank-order operator is the overall procedure by which objects or processes being measured are ordered and assigned values of the quantity being measured. In length-based dimensional measurements, the ordering operator is equivalent to a comparison of two objects involving, in effect, a translation in physical space. Each type of dimensional measurement, however, requires its own particular realization of that ordering operator. As a result, displacement, position, distance, and extension each constitutes its own measurement dimension which must be operationally tied to the dimension of physical space.

b. *The Unit:* The *unit* is the rank-order greater of two objects or states of the physical system assigned a defining numerical value of the quantity being measured. The unit is one of the two points required to specify what is axiomatically a linear system. In length-based dimensional measurements which conform to international standards, the unit is the SI unit of length. In practical dimensional measurements today, the unit of length is an internationally-accepted value for a well-defined wavelength of an iodine-stabilized HeNe laser as discussed above in Section IIIA.

c. *The Zero:* The *zero* is the rank-order lesser of the two objects or states of the physical system assigned a defining numerical value of the quantity being measured. In effect, the zero is the other of the two points required to specify the linear system. Implied by the definition of the unit of length is that the *zero* of measurement has an assigned numerical value of zero (that is, $| \ell_z |$ $\equiv 0$) and that the zero of length of dimensional measurements is the length of the path in space which light traverses in the zero interval of time.

In practical dimensional measurements, the *zero* corresponds to the physical zero-vector origin of coordinates which must be practically realized in each of the different dimensional types. For example, in laser interferometry, the zero of displacement is the initial optical-path difference between the reference beam and moving-mirror beam, one which must remain constant during subsequent translation of the mirror in order to prevent loss of origin of displacement. In coordinate measuring machines, the zero of position is the initial location of the tip of the probe relative to the interferometer reflector, which must remain constant as the probe is moved about to prevent loss of origin of position.

d. *The Scale:* The scale is the set of equal graduations into which the difference between the unit and the zero is divided. In effect, the scale is the condition of linearity under which *differences* in dimensions are equal, that is, the difference between one pair of objects is the same as the difference between another pair. Operationally, the scale is the means for measuring in increments smaller than the unit.

In practical dimensional measurements, the *scale* is the combination of a physical transducer and a linearization algorithm by which linear interpolations between known points can be carried out. For example, in displacement interferometry, the scale is generated by interpolating between fringes using an assumed sinusoidal function of waveforms for the linearization algorithm. Non-linear transducers such as capacitance gauges -- when used in conjunction with suitable mathematical models of their responses -- can also be used to generate the required linear scale.

$$\bullet \ \bullet \ \bullet$$

In sum, the axioms of measurement define the requisite elements of a measurement system, each of which must be operationally realized to achieve meaningful results. The first axiom — the rank-order operation — defines the dimension of measurement, such as temperature, time, voltage, or, in this case, each of the four types of length. The other three axioms are the means by which numerical values are associated with measurements of that dimension. Table 5 below shows explicitly the form of the axiom errors for each of the dimensional types.

Table 5.   The Matrix of Errors by Measurement Axiom and Dimensional Type in General Form

Dimensional Type	Error of the Zero	Error of the Unit	Error of the Scale
Displacement	a_D	$b_D \cdot D$	$c_D(D^n)$
Position	a_P	$b_P \cdot P$	$c_P(P^n)$
Distance	a_d	$b_d \cdot d$	$c_d(d^n)$
Extension	a_e	$b_e \cdot e$	$c_e(e^n)$

2. Analysis of Error in Terms of the Axioms

Associated with each of the latter three axioms of measurement is a specific type of error with specific, fixable causes. Errors of the unit, the zero and the scale correspond respectively to the terms in Eq. 26 with *a* equal to the error of the zero, b to the error of the unit, and $c(\ell_t^n)$ to the error of the scale, that is, the sum of all non-linear terms.

a. *Error of the Zero*: Errors of the zero occur for measurement systems which have a properly linear relation of measured value to true value, ℓ_{obs} to ℓ_t, but have a non-zero value at $\ell_t = 0$, that is:

$$\ell_{obs} = \ell_t + a \tag{27}$$

and $\qquad \Delta \ell_z = a.$

An error of the zero arises, for example, due to a shift in the origin of coordinates in the course of a dimensional measurement.

b. *Error of the Unit*: Errors of the unit occur for measurement systems which — while having a properly linear relation of measured value to true value — have a non-unitary slope, that is:

$$\ell_{obs} = (1+b) \cdot \ell_t \tag{28}$$

and $\qquad \Delta \ell_u = b \cdot \ell_t.$

Error of the unit corresponds, for example, to an erroneous value of wavelength in displacement interferometry or a constant misalignment of coordinate and object axes.

c. *Error of the Scale*: Errors of the scale occur for measurement systems which have a non-linear relation of measured value to true value, that is:

$$\Delta \ell_s = c(\ell_t^n). \tag{29}$$

Error of the scale represents failure of the interpolation scheme to divide the difference between the unit and the zero into strictly equal intervals, that is, the degree to which the system is nonlinear. The measurement system is linear, if given four objects (i.e., four displacements, positions, distances or extensions) for which the true difference between pairs is the same, the measured differences are also the same. Error of the scale is, by definition, the inequality in those differences, that is:

$$\Delta_s = (\ell_{obs1} - \ell_{obs2}) - (\ell_{obs3} - \ell_{obs4}) \tag{30}$$

when $\qquad (\ell_{t1} - \ell_{t2}) = (\ell_{t3} - \ell_{t2}) .$

That the error of the scale corresponds to a non-zero value of the coefficient $c(\ell_t^n)$ can be seen when the non-linear response of a dimensional measurement system is given a specific form, such as the quadratic:

$$\ell_{obs} = a + (1+b) \cdot \ell_t + c\ell_t^2 . \tag{31}$$

Evaluation of Eq.30 at four values of ℓ_t, denoting the differences in the pairs of true values by Δ_i, and substitution into Eq. 26 leads to the result:

$$\Delta \ell_s = c \cdot (\ell_{t1} - \ell_{t3}) \cdot \Delta_i . \tag{32}$$

An error of the scale arises, for example, from to faulty interpolation of fringes in interferometry or uncompensated nonlinearities in LVDT or capacitive displacement measuring devices.

d. *Errors of Zero, Unit and Scale*: In summary, the axiom-specific errors for the general case that $\ell_{obs} = a + (1+b) \cdot \ell_t + c(\ell_t^n)$ are given by:

$$\Delta \ell = \Delta \ell_z + \Delta \ell_u + \Delta \ell_s = a + b \cdot \ell_t + c(\ell_t^n) \tag{33}$$

where: a corresponds to $\Delta \ell_z$, the error in realizing the zero; b corresponds to $\Delta \ell_u$, the error in realizing the unit; c corresponds to $\Delta \ell_s$, the error in realizing the scale; and ℓ is any one of the four dimensional measurement types.

V. PRINCIPAL SPECIFIC SOURCES OF ERRORS

Errors in dimensional measurements are particular to each of the different types and, since each succeeding type is dependent on that which precedes it, total error increases as one proceeds from the displacement, position and distance to the extension of objects.

A. Error in Displacement Interferometry

The type of dimensional measurement most directly linked to the SI unit of length and the basis for all subsequent types of dimensional measurements is displacement by laser interferometry, that is, the measurement of a linear change of location in space with time of a single object, in this case, the moving mirror of the interferometer. The measure of the limit of this ability is displacement uncertainty or error, δD. Errors in measurement of displacement by laser interferometry are associated with each of the terms in Eq. 15.

1. Error of Wavelength in Vacuum λ_o

The uncertainty in the wavelength in vacuum of the laser light depends on the type of stabilization used in its construction. As indicated in Table 3 above, uncertainty in the vacuum wavelength of the 633-nm red-orange line of the helium-neon (HeNe) lasers widely used in dimensional metrology range from 10^{-6} for unstabilized lasers down to 10^{-10} for iodine-stabilized ones.

2. Error of Index of Refraction of Medium n

The uncertainty in the wavelength in medium of the laser light depends on the index of refraction of the medium through which the light propagating in the interferometer passes.

a. *Error of the Index of Ambient Air:* Since index of refraction of a gas is a function of its temperature, pressure, humidity and composition, without compensation for actual variations in those parameters the upper limit of uncertainty for interferometric displacement measurements in ordinary ambient air can be large: for example, an index error of 10^{-6} would result from any one of the following variations: a one degree Centigrade change in temperature, a 2.5 mm Hg change in atmospheric pressure, or an 80% change in relative humidity [13].

b. *Error of the Index of Compensated Standard Air:* Errors associated with the index of refraction of the ambient medium can be reduced by compensation achieved by measurement of the index of the actual ambient air or by calculation of the index of standard dry air. The Edlen formula is an internationally agreed upon equation for the calculation of the index of refraction of standard air as a function of wavelength, air temperature, air pressure and relative humidity [14].

With compensation, the lower limit of uncertainty for practical laser displacement interferometry in ambient air is estimated to be about $1.2 \cdot 10^{-7}$ [15]. The uncertainty in the Edlen formula itself for the index of refraction of standard dry air is estimated to be $5 \cdot 10^{-8}$ [16]. Operation of an interferometer in vacuum eliminates this source of error.

c. *Error of Index of Refraction in the Dead Path:* So-called "dead path" error is due to improper compensation for the difference in the lengths of the optical paths of the two interfering beams at the zero-displacement position of the system when index-altering environmental changes occur during the course of the displacement measurement. For commercial systems which use sensor-based index compensation, dead-path errors are largely compensated for in software, leaving a residual error of $1.4 \cdot 10^{-7}$ times the dead-path distance.

3. Error in Fringe-Fractioning

Errors in realizing a scale by means of displacement interferometry arise from various sources in the process of the generation, subdivision and counting of the interference fringes. Representative of such errors are those which arise in widely-used displacement interferometers of the polarization beam-splitting type. The principal sources of error in this type of system are those associated with electronic subdivision, polarization mixing, thermal drift and dead path [17].

a. *Error in Electronic Subdivision*: The inherent half-wavelength resolution of interferometers, corresponding to the spacing of the alternating light and dark of the fringes, can be extended by a variety of electronic fringe-interpolation schemes, each of which has its own limiting resolution which contributes an additive, least-count error. Operating at 633 nm, certain commercial polarization heterodyne interferometer systems have least counts in electronic-subdivision, depending on the whether used with retro-reflectors or plane mirrors, of $\lambda/32$ (approximately 20 nm) and $\lambda/64$ (approximately 10 nm) respectively.

b. *Error from Polarization Mixing*: Less-than-perfect separation of the polarization states of the interfering beams in polarization-type interferometers, due to leakage of one component into the other, produces a non-linear error in displacement measurement with such systems. Such error varies as a function of change in optical path with a periodicity of the wavelength of the laser source with an amplitude specific to an individual interferometers. For one commercial linear interferometer system, the peak-to-peak phase error was found to be 5.4° corresponding to approximately 5 nm.

c. *Error from Thermal Drift*: Changes in temperature within the optical components of the interferometer system, which produce a differential change in path optical length between the interfering beams, give rise to drift type errors in displacement measurement. This environmentally induced error is less for temperature-controlled systems and those which are specifically designed to deal with this source of error. An example of the latter has a quoted thermal error of 40 nm/C°, twelve times better than a conventional plane-mirror system with its 0.5 μm/C° [12].

4. Error in Alignment of Interferometer

Finally, constant angular misalignment between the interferometer and the incident light gives rise to an error (derived from the basic interferometer equation):

$$\delta D_\theta = D \cdot (-\theta^2/2), \qquad (34)$$

where δD_θ is misalignment contribution to the displacement error, D is the measured displacement, and $\delta\theta$ is the small angle of misalignment between the axis of interferometer (the normal to the parallel faces of the interferometer mirrors) and the axis of propogation of the incident laser light.

· · ·

In sum, this section has looked at sources of errors in the type of dimensional measurement most directly linked to the SI unit of length, those of laser displacement interferometry by the best available form of that technique, polarization-type heterodyne interferometry.

B. Error in Coordinate-Position Measurement

After displacement, the type of dimensional measurement next most directly linked to the SI unit of length and inherent in subsequent types is position, that is, measurement of the location of a single point relative to a coordinate system. The measure of the limit of the ability to measure position is position error, δP. Errors in position measurements are associated with each of the functional components of the coordinate measurement system as embodied by a coordinate measuring machine: the scales, frame, carriage and probe.

1. Error In Relation to Displacement Scales

Lack of commonality of origin and co-linearity of axes of the coordinate-position measurement system and the displacement system, which supplies the metric and scale subdivisions, gives rise to two types of errors specific to positional measurement.

a. *Misalignment of Mirror Translation*: Constant angular misalignment between the axis of the displacement measurement and the axis of the coordinate system gives rise to "cosine error," which is governed by the relationship:

$$P = D / \cos \alpha \tag{35}$$

where D is the displacement, P is the coordinate position being assigned by the measurement and α is the angle of misalignment between the two. For a displacement generated by an interferometer system, α is the angle between the axis of translation of the moving mirror and the optical axis of the interferometer (i.e., the normal to the parallel faces of the reference and moving mirrors). For a small angle, the cosine error contribution to coordinate-position error, a multiplicative error, is given by:

$$\delta P_c = D \cdot (-\alpha^2/2). \tag{36}$$

Cosine error arises, in effect, when the axis of the displacement vector and axis of the coordinate vector are rotated with respect to each other but have a common origin. [Note that the errors due respectivelt to misalignment of interferometer and light (Eq.34) and misalignment of mirror translation and interferometer (Eq.36) are of the same sign while that due to misalignment of object and mirror translation (Eq.40) is opposite (Appendix A)].

b. *Abbe Error:* Abbe error can arise when the axes of the displacement vector and the coordinate vector do not share a common origin *and* they rotate with respect to each other, that is, tilt, during the course of the displacement. Abbe error is governed by the relationship:

$$D = P + O_a \cdot \sin \phi \qquad (37)$$

where D is the measured displacement, P is the coordinate being assigned by the measurement, O_a is the lateral distance between the displacement and coordinate axes (called the Abbe offset), and ϕ is the angle of tilt which occurs during the course of the displacement measurement. For small angles, the Abbe error contribution to coordinate-position error, which is an additive, is given by:

$$\delta P_a = O_a \cdot \phi. \qquad (38)$$

2. Error Related to Carriage Motion

Carriage-associated position-measurement error occurs in measuring machines in which the motion-generating system (which translates the object relative to the reference frame) is the same as the displacement-measuring system (which measures that translation) as, for example, in a machine which uses the same lead screw both to move a carriage and to measure its position. For such machines, carriage-associated errors include backlash and hysteresis, both of which are direction-dependent and result in loss of origin of the coordinate system.

3. Errors Related to Reference-Frame Geometry

Limitations in the mechanical structure of a coordinate measuring machine give rise to positional errors associated with the machine's coordinate axes and their angular relationships. Since the reference frame of a coordinate measuring machine must consist of one highly-planar reference surface for each coordinate axis with each of these planes angularly oriented with respect to the others in a constant and well known way, the errors associated with reference frame geometry are complex.

Characterization of the positional errors in a 3D CMM requires dealing with twenty-one degrees of freedom based on six degrees per axis (three translational plus three angular, including roll, pitch and yaw) plus the three angular orientations of the axes with respect to each other. Specialized matrix-based models and measurement algorithms for characterizing positional errors in coordinate measuring machines and machines tools have been developed [18,19] and standardized techniques adopted [20].

4. Error Related to the Probe

Limitations in the mechanism for linking the object to be measured to the machine's coordinate measurement system give rise to positional errors specifically associated with the machine's probe. Probe-specific errors are of two types, depending on whether the probe is designed to carry out only functions of a probe as a simple sensor or a probe as a measuring machine.

a. *Positional Error of Probe-as-Sensor*: Positional errors of a probe-as-sensor arise to the degree that it fails to carry out its primary function, that of detecting -- in a binary sense of "here" or "not here" -- the boundary of an object, such as to locate it relative to the coordinate system of the machine. Positional errors of probe-as-sensor include what in touch-trigger mechanical-contact probes is called "pre-travel," that is, the displacement of the probe tip after contact with the object but before the probe signals that contact. Such error arises, for example, from variations in bending of the probe structure and the threshold response of the mechanical-to-electrical transducer.

b. *Positional Error of Probe-as-Measuring-Machine*: Positional errors of a probe-as-measuring-machine arise in, for example, "analog probes," which are designed not only to carry out the primary function of linking machine and object, but also to provide an output signal proportional to the displacement of the probe tip on contact with the object. Positional errors of analog probes, being as they are miniature measuring machines attached to a larger machine include: 1) errors of the probe as a sensor; 2) errors of the probe as a measuring machine (including those associated with scale, frame, and carriage); and 3) errors associated with the orientation of the probe as one coordinate system with respect to another.

• • •

In sum, this section has looked at sources of errors in measurements of position, which after displacement, is the next most directly linked to the SI unit of length. As indicated, position error deals with the ability to measure location of points in space and are associated with each of the functional components of a coordinate measuring machine by which such position measurements are made.

C. Error in Distance Measurement

After displacement and position, the type of dimensional measurement next most directly linked to the SI unit of length is that of distance, that is, measurement of the separation of two successively located point-like features on material objects. The measure of the limit of the ability to measure distance is distance error, δd. Errors specific to measurements of distance are associated with the probing, angular alignment and environmental-dependence of separation of distance-type features of the object.

1. Distance Error from Feature Probing

Errors of feature probing arise from variations in the size, shape and material of the individual point-, line- and plane-like features, the separation of which is being measured in a distance measurement. Such variations limit the ability to reproducibly locate features with a probe. Errors associated with such variations are assessed in terms of the repeatability with which the probe of a particular coordinate measuring machine can be set on a particular feature.

2. Misalignment of Object and Coordinate Axes

Misalignment between the object being measured and a coordinate axis of the measuring machine gives rise to cosine error governed by the relationship:

$$d = |\, P_1 - P_2 \,| \cdot \cos \gamma, \tag{39}$$

where P_1 and P_2 are the measured coordinate positions of the point-like features, d is the distance being measured and γ is the angle of misalignment between the coordinate axis and that of the object itself. For a small angle, the cosine error contribution to distance error is given by:

$$\delta d_c = d \cdot (+\gamma^2/2). \tag{40}$$

3. Error from Object Environment

Apart from the effects of changes in the environment on displacement-measuring devices and the measuring machine itself, such environmental changes give rise to errors associated with the object itself, including those associated with thermal expansion, mechanical distortion and contamination.

a. *Error from Object Thermal-Expansion*:   Distance error due to thermal expansion of the object can be due to different temperatures for the whole object or different temperatures for two positions on the object. In either case, variation in the dimensions of the object with temperature are governed by the equation:

$$\rho = - (\Delta \ell / \ell_o) / \Delta T, \tag{41}$$

where ρ is the coefficient of linear thermal expansion of the material and $\Delta \ell / \ell_o$ is the fractional change in the "length" of material for a difference in temperature ΔT. Note that since thermal expansion is defined in terms of the *change* in length (either between distance-type features or extension-type boundaries), it is inherently a distance-type property given by:

$$\delta d_T = \rho \cdot \Delta T. \tag{42}$$

b. *Error from Object Distortion*: Distance error can also arise from any other factor which, in effect, distorts the object by altering the relative separation of features of the object. Such factors include any local or overall mechanical distortion of the object due, for example, to fixturing of the object on the measuring machine; and any gain or loss of material on one feature with respect to the second due, for example, to contamination or wear.

• • •

In sum, this section has looked at sources of errors in measurements of distance is, which after displacement and position, the next most directly linked to the SI unit of length, such errors being associated with measurement of the separation of two point-like features of material objects.

D. Error in Extension Measurement

The type of dimensional measurements operationally farthest removed from the primary realization of the SI unit of length is that of extension. The measure of the limit of the ability to measure extension is extension error, δe. Most specific to the measurement of extension-type features of objects, such as the inside diameters of holes and the outside diameters of plugs, is error associated with locating a single boundary and error associated with locating one boundary relative to another one.

1. Error in Location of a Single Boundary

a. *Error in Reducing a 3D Surface to a Point*: Errors occur in measuring the location of a real boundary because a material object is not an ideal geometrical surface being intersected by a line such as to define a point. Instead that boundary is a rough three-dimensional surface being contacted by a finite-area probe. Estimates of the contribution to boundary-location errors for specific object-probe combinations may be made based on measurements of surface roughness, which involves statistical averaging over surface irregularities. Boundary-location errors due to surface roughness can be comparable to the roughness itself, which for a highly-polished steel surface, for example, can be of the order of 0.05 μm.

b. *Error in Compensation of Probe-Object Interaction*: Errors occur in measuring the position of a single boundary as a result of any uncompensated off-set between the location of the probe and the location of the surface which the probe signals. For example, in locating the boundary of a machined part with a mechanical stylus, single-boundary-location errors occur due to incomplete compensation for both: the finite radius of the probe and its finite penetration into the part. Analogous errors occur in locating the boundary of a microcircuit element with the beam of an electron microscope.

Compensation for probe-object interaction, such as the deformation of a part by a mechanical stylus and penetration into the bulk of a material by an electron beam, can only be done by fundamental theoretical modelling of the specific probe and object involved. Of necessity, such modelling has been done, for example, for high-accuracy optical- and electron-microscope calibrations of photomask linewidth standards for use in the microelectronics industry [21].

$$\bullet \ \bullet \ \bullet$$

In sum, error in boundary location is a vector having a magnitude (associated with, for example, the finite roughness of the surface, the finite extent of the probe and the finite depth of penetration of the object by the probe) and a direction, positive or negative, relative to the axis of coordinate measurement.

2. Error in Location of Two Boundaries

In measurements of extension, errors in the location of one boundary combine as vectors with errors in the location of the second boundary — that is, one of opposite orientation — and the results are non-canceling, as indicated by:

$$\delta_e = \delta P_{b1} + \delta P_{b2}, \tag{43}$$

where δP_{b1} is the error in the location of the first boundary and δP_{b2} is the error in the location of the second.

For the case that the boundary-location error is the same for the two boundaries, that is, where the magnitude of δP_{b1} equals that of δP_{b2}, the error in extension measurement due to the combination of the two boundary-location errors is:

$$\delta_e = \mp 2\delta P_b, \tag{44}$$

where the minus sign is for outside-caliper measurements (as in the diameter of a plug) and the plus sign is for inside-caliper measurements (as in the diameter of a hole) when the boundary-location errors are due, in effect, to "penetration" (positive or negative) of the object by the probe.

An example of a positive boundary-location error is that due to the finite radius of a mechanical probe (that is, the extension of the probe tip along the axis of measurement). An example of a negative boundary-location error is probe pre-travel (that is, the finite displacement of the probe along the axis of measurement after mechanical contact with the object, but before the probe transducer crosses a threshold and signals).

Additive boundary-location errors are unique to extension-type measurements. In measurements of displacement, position and distance, the boundaries involved face in the same direction and boundary-location errors cancel, that is:

$$\delta P_{D,P,or\ d} = \delta P_b - \delta P_b \equiv 0. \tag{45}$$

E. Compounding of Error in Measurement Types

A final aspect of assessing accuracy in practical dimensional measurements involves a look at the cascading nature of the errors in the succession of the dimensional measurement types. This cascading, that is, accumulation, of errors is shown formally below, where Δ represents the total uncertainty of the type of measurement and δ represents the type-specific error in that total.

In this notation, the total error in displacement measurement ΔD a function of δD, which is all the displacement-specific errors described in Section IVA, plus $\Delta\lambda_o$, which is the error specific to the metric-carrying vacuum wavelength of laser light source, and is given by:

$$\Delta D = \Delta D(\delta D, \delta\lambda_o). \tag{46}$$

Similarly, the total error in position measurement Δp a function of δp, which is all the position-specific errors described in Section IVB, plus ΔD, which is the total error in displacement measurement, and is given by:

$$\Delta p = \Delta p(\delta p, \Delta D) \tag{47}$$

$$= \Delta p(\delta p, \delta D, \delta\lambda_o).$$

And again, the total error in distance measurement, Δd, a function of the total error in position measurement, plus all the distance-specific errors described in Section IVC above, and is given by:

$$\Delta d = \Delta d(\delta d, \Delta p) \tag{48}$$

$$= \Delta d(\delta d, \delta p, \delta D, \delta\lambda_o)$$

Finally, the total error in extension measurement, Δe, a function of the total error in distance measurement, plus the extension-specific errors described in Section IVD above, and is given by:

$$\Delta e = \Delta e(\delta e, \Delta d) \tag{49}$$

$$= \Delta e(\delta e, \delta d, \delta p, \delta D, \delta\lambda_o)$$

As is indicated by this formal analysis, the accuracy of each type of dimensional measurement depends on — and is limited by — the total error of the preceding type, as well as by the sum of its own type-specific errors.

• • •

In sum, measurements of extension are the most difficult in which to achieve accuracy because they inherit all the errors of the successive displacement, location and distance measurements upon which they are based and, addition, are subject to intractable boundary-location errors to which the others are not.

VI. ERROR-BUDGET EXAMPLE

All of the considerations of accuracy assessment discussed above can be taken into systematic account in the assessment of the errors in a particular dimensional measurement system by compilation of an error budget for the measuring system in question. In this example, a simple sum-over-errors is used to estimate the total worst-case error; other means for estimating the most likely errors have been described [22].

The example considered here is measurement of an extension-type object on a laser-interferometer-based single-axis coordinate measuring machine. In this example, the overall system a function of a stabilized HeNe laser, a commercial heterodyne interferometer system, a machine frame with a moving carriage, a touch-fire probe, and a 100mm-long steel rod, the length of which is to be measured, all within a normal atmospheric environment.

Table 6 below shows a compilation of the contributions to the overall accuracy of measurement of the specified object with the hypothetical coordinate-measuring-machine system described. The organization of the error budget follows the sequence of the type scheme used throughout the chapter. In the table, $\delta\ell_{Ai}$ and $\delta\ell_{Bi}$ represent, respectively, additive and multiplicative contributions to the type-specific errors. The values are estimates of the relative size of error contributions from the various sources. In the following sections, each of the individual contributions to the type-specific errors for this hypothetical example are discussed in turn.

A. Displacement Errors

In this example the errors in the laser-interferometric measurement of the displacement derive from displacement-specific errors associated with each of the parameters in the basic interferometer equation plus the error associated with the vacuum wavelength of the laser light.

Table 6.   **Error-Budget Example: An Analysis of a Measurement of the Extension of a 100mm-Long Object with an Interferometer-Based Coordinate Measuring Machine**

Error Contribution by Source		$\delta \ell_{Ai}$	ℓ_{Bi}
Interferometer Least Count	δN_{lc}	0.005μm	
Beam-Polarization Mixing	δN_{pm}	0.002μm	
Machine-Path Atmosphere	δn_{pa}		10 ppm
Compensated	δn_{pc}		0.15 ppm
Optics Thermal Drift	δn_{io}	0.02 μm	
Dead-Path Compensated	δn_{dp}	0.015μm	
D-vs-λ Alignment	$\delta \theta_{to}$		0.05 ppm
Sum of D-Specific Errors	δD	0.042μm	0.20 ppm
Vacuum-Wavelength Error	$\delta \lambda_o$		0.02 ppm
<u>Displacement Error Total</u>	ΔD	<u>0.042μm +</u>	<u>0.22 ppm</u>
P-vs-D Alignment	δP_ϕ		0.05 ppm
Carriage Abbe Tilt	δP_β	0.24 μm	
Probe Setting	δP_p	0.50 μm	
Sum of p-Specific Errors	δP	0.74 μm	0.05 ppm
Displacement Error Total	ΔD	0.042μm	0.22 ppm
<u>Position Error Total</u>	Δp	<u>0.782μm +</u>	<u>0.27 ppm</u>
O-vs-P Alignment	δd_γ		1.5 ppm
Thermal Expansion	δd_ρ		1.0 ppm
Point-Feature Definition	δd_s	0.10 μm	
Sum of d-Specific Errors	δd	0.10 μm	2.5 ppm
Position Error Total	Δp	0.782μm	0.27 ppm
<u>Distance Error Total</u>	Δd	<u>0.882μm +</u>	<u>2.77 ppm</u>
Arbitrarily Located Bndry	δe_{bx}	1.0 μm	
Compensated	δe_{bt}	0.1 μm	
Shape-Averaged Boundary	δe_{bm}	0.2 μm	
Sum of e-Specific Errors	δe	0.3 μm	
Distance Error	Δd	0.882μm	2.77 ppm
<u>Extension Error Total</u>	Δe	<u>1.182μm +</u>	<u>2.77 ppm</u>
Error in e @ 100mm	Δe_{100}	1.459μm →	±1.5 μm

1. Fringe-Fraction Error δD_N

The contributions to displacement-specific measurement error are due to errors in determination of the change in the order of interference with the translation of the interferometer reflector:

$$N = i + f, \tag{50}$$

where N is a real number which is the sum of the counted order i and measured fringe fraction f. Practically, the total error in N is associated with error in interpolating between fringes, that is:

$$\delta N = \delta f. \tag{51}$$

In this example, the contributors to error in fringe-fractioning are the 5nm due to the least count of interferometer and the 2 nm due to the mixing of the polarization states of the interfering beams.

2. Index-of-Refraction Error δD_n

Contributions to displacement-specific measurement error arise from variations in the indices of refraction along the optical paths of the interfering beams due to changes in environmental conditions. The variation in phase between the two beams as a function of variations in the individual indices of refractions in each of the segments of the reference- and moving-arm beams is given by:

$$\delta \Delta \phi = (2\pi/\lambda_o) [\Sigma_i \, \delta n_{ri} \cdot \ell_{ri} - \Sigma_i \, \delta n_{mi} \cdot \ell_{mi}], \tag{52}$$

where Σ_i indicates a sum over the i number of optical-path segments $n_i \cdot \ell_i$ in each of the arms. In this example, the principal segments of the overall optical path include the CMM machine-axis path along which the moveable reflector is translated, the dead path, and the interferometer-optics path, that is, the glass through which the light passes.

a. *Machine-Path Atmosphere*: The first and major source of index-of-refraction errors is the variation of the atmospheric air along the displacement path. The following show the changes in temperature, pressure, and relative humidity each of which gives rise to a displacement error of 10^{-6}:

$$\delta n_{\Delta T} = 1 \text{ ppm} / 1°C$$
$$\delta n_{\Delta P} = 1 \text{ ppm} / 2.5 \text{ mm Hg} \tag{53}$$
$$\delta n_{\Delta H} = 1 \text{ ppm} / 80\% \, \Delta RH.$$

In this example, a $\pm 0.5°C$ change in temperature, a ± 25 mm change in pressure, and a $\pm 10\%$ change in relative humidity combine to produce a multiplicative error of about 10 ppm. With measurement of the changes and compensation for them, this is reduced to 0.15 ppm.

b. *Interferometer-Optics Thermal Drift*: The second source of index of refraction error is change in the temperature of the glass components of the interferometer through which the light in the two beams passes. In this example, optics thermal drift error is 0.02 μm.

c. *Dead-Path Compensation*: A last source of index of refraction error is change in temperature, pressure, and humidity of the dead-path (the difference in the interferometer arms at the starting point of the machine-axis path). In this example, compensation has reduced dead-path error to 0.015 μm.

3. Interferometer-Axis Alignment Error $\delta D_{\theta ti}$

The contribution to displacement-specific measurement error due to the angular misalignment θ_{ti} between the axis of the interferometer and the axis of propagation of the incident light is, for a small angle of misalignment, given by:

$$\delta D_{\theta ti} = D \cdot (- \theta_{ti}^2/2). \tag{54}$$

For an angular alignment corresponding to a lateral displacement of one-tenth the spot size of a 3mm-beam at the one-meter travel of the measuring machine stage, $\theta_{ti} = 3\times10^{-3}$ radians and the resulting error $\delta D_{\theta ti}$ is 0. 05 ppm.

4. Sum of Displacement-Specific Errors δD

In this example, the total displacement-specific error δD is the sum of the errors due to fringe-fractioning, index of refraction, and angular misalignment of translation-propagation axes and is given by:

$$\delta D = \Sigma_i \, \delta D_i$$

$$= (\Sigma \delta D_{Ai}) + (\Sigma \delta D_{Bi}) \cdot \ell \tag{55}$$

$$= 0.042 \, \mu m + 0.20 \, ppm \cdot D \, .$$

At a displacement of 100 mm (corresponding to the extension of the object being measured), the sum of the displacement-specific errors is:

$$\delta D_{100} = 0.062 \, \mu m = 0.62 \, ppm \, @ \, 100 \, mm \tag{56}$$

5. Vacuum-Wavelength Error $\delta\lambda_o$

The contribution to the total displacement measurement error ΔD due to the error in the value of the vacuum wavelength of the laser is:

$$\delta D_{\lambda o} = (\delta\lambda_o / \lambda_o) \cdot D, \tag{57}$$

which for the highly stabilized commercial laser of this example is $2 \cdot 10^{-8}$. Thus,

$$\delta D_{\lambda o} = 0.02 \text{ ppm} \cdot D. \tag{58}$$

6. Total Displacement Error $\Delta D(\delta D, \delta\lambda_o)$

As indicated, combining the displacement-specific error δD with the error in the vacuum-wavelength of the laser source $\delta\lambda_o$ yields the total displacement error ΔD over the one-meter travel of the measuring machine stage, that is:

$$
\begin{aligned}
\Delta D &= \delta D + \delta\lambda_o \\
&= (\Sigma\delta D_{Ai}) + (\Sigma\delta D_{Bi}) \cdot \ell + (\delta\lambda_o/\lambda_o) \cdot \ell \qquad (59) \\
&= 0.042 \text{ }\mu\text{m} + 0.22 \text{ ppm} \cdot D
\end{aligned}
$$

For a displacement measurement corresponding to length of the 100-mm-long object, the total error in the displacement measurement is:

$$\Delta D = 0.064 \text{ }\mu\text{m} = 0.64 \text{ ppm} @ 100 \text{ mm.} \tag{60}$$

B. Error in Position Measurement ΔP

In this example the errors in the coordinate-measuring-machine measurement of position derive from errors associated with the geometrical relationships among the scale, frame, carriage and probe of the CMM and the setting ability of the probe itself.

1. Misalignment of Coordinate and Displacement

The contribution to position-specific measurement error due to the angular misalignment ϕ between the axis of translation of the mirror and the axis of the interferometer is, for a small angle, given by:

$$\delta P_\phi = D \cdot (- \phi^2/2). \tag{61}$$

For an angular misalignment of $\phi_{ti} = 3 \cdot 10^{-3}$ radians, the resulting error is:

$$\delta P_\phi = 0.05 \text{ ppm} \cdot P. \tag{62}$$

2. Carriage Abbe Tilt

The contribution to position-specific measurement error due to a difference in angle of tilt of a non-zero-length probe between two positions (Abbe error) is, for a small angle:

$$\delta P_a = O_a \cdot \beta. \tag{63}$$

In this example, for a distance between the axis of the probe and the axis of the displacement measurement of 50 mm and for a change in tilt of 1 arc second, the Abbe error contribution to coordinate-position error, which is an additive, is:

$$\delta P_a = 0.24 \ \mu m. \tag{64}$$

3. Probe Setting

The contribution to position-specific measurement error due to probe setting is a measure of the ability of an individual probe to reproducibly locate unvarying, uni-directionally approached, probed features.   Since a location measurement requires the probe to trigger twice (once at the location of the origin and once at the location being measured, an estimate of the error in probe-setting based on the 2σ setting of a good-quality commercial probe operated in the touch-trigger mode is:

$$\delta P_p = 2 \cdot 0.25 \ \mu m = 0.50 \ \mu m. \tag{65}$$

4. Sum of Position-Specific Errors δp

In this example, the total position-specific error δP is the sum of the errors due to coordinate-scale misalignment, Abbe error, and probe setting.

$$\delta P = \Sigma \ \delta P_i$$

$$= 0.74 \ \mu m + 0.05 \text{ ppm} \cdot d \tag{66}$$

$$= 0.745 \ \mu m \ @ \ 100mm$$

5. Total Position Error $\Delta P(\delta P, \Delta D)$

Combining the position-specific errors, δp, with the total displacement error, ΔD, over the one-meter travel of the measuring machine stage yields:

$$\Delta P = \delta P + \Delta D = 0.782 \ \mu m + 0.27 \ ppm \cdot P. \qquad (67)$$

For a position corresponding to length of the 100mm-long object, the total error in the position measurement is:

$$\Delta P_{100} = 0.809 \ \mu m = 8.1 \ ppm \ @ \ 100 \ mm \qquad (68)$$

C. Error in Distance Measurement Δd

In this example the errors in the coordinate-measuring-machine measurement of distance derive from distance-specific errors associated with the object alignment, temperature and distance-type features, plus the error in position measurement.

1. Object-Axis/Probe-Path Alignment

The contribution to distance-specific error due to the angular misalignment γ between the axis of the object and the axis of coordinate measurement, that is, the path of the probe, is, a small angle, given by:

$$\delta d_\gamma = d \cdot (+ \gamma^2/2). \qquad (69)$$

For an angular misalignment of $0.1°$ (corresponding to a lateral distance of less than 0.2 mm over the 100mm length of the object):

$$\delta d_\gamma = 1.5 \ ppm \cdot d . \qquad (70)$$

Note that this error due to misalignment of object and coordinate axis (Eq.40) is opposite in sign to that due to misalignment of interferometer and light (Eq.34) and misalignment of mirror translation and interferometer (Eq. 36).

2. Object Thermal Expansion

The contribution to distance-specific error due to the difference in temperature ΔT between the object to be measured and a reference temperature (such as that of the CMM) is determined by the linear thermal expansion coefficient ρ for the object according to the thermal expansion equation:

$$\delta d_{\Delta T} = \rho \cdot \Delta T \cdot d \qquad (71)$$

For the a steel object in this example, ρ is 10 ppm/$°C$, which for temperature difference ΔT of $0.1°C$ gives:

$$\delta d_{\Delta T} = 1.0 \ ppm \ x \ d. \qquad (72)$$

3. Object Point Definition

The contribution to distance-specific measurement error due to object point definition is a measure of the ability to associate with the object features which can be characterized as points, the position of which points can be reproducibly located by a variability-free probe. In this example, the estimate of a practical lower limit for error associated with locating points on real objects (twice that for defining a single point) is estimated to be:

$$\delta d_s \quad = 0.10 \; \mu m. \tag{73}$$

4. Sum of Distance-Specific Errors δd

In this example, the total distance-specific error δD is the sum of the errors due to coordinate-scale misalignment, Abbe error, and probe setting.

$$\delta d \quad = \Sigma \; \delta d_i \quad = \quad 0.10 \; \mu m + 2.5 \; ppm \cdot d. \tag{74}$$

At a distance of 100mm corresponding to the extension of the object being measured, the sum of the distance-specific errors is:

$$\delta d_{100} = 0.35 \; \mu m = 3.5 \; ppm \; @ \; 100 \; mm. \tag{75}$$

6. Total Distance Error $\Delta d(\delta d, \Delta p)$

Combining the sum of the distance-specific errors, δd, with the total position error, Δp, over the one-meter travel of the measuring machine stage yields:

$$\Delta d \quad = \delta d + \Delta p \quad = \quad 0.882 \; \mu m + 2.77 \; ppm \; x \; d. \tag{76}$$

For a distance corresponding to the length of the 100-mm-long object, the total error in the distance measurement is:

$$\Delta d \quad = 1.109 \; \mu m = 1.1 \; ppm \; @ \; 100 \; mm. \tag{77}$$

D. Error in Extension Measurement Δe

In this example, the errors in the coordinate-measuring-machine measurement of extension derive from extension-specific errors associated with boundary location and definition, plus the error in distance measurement.

1. Boundary-Location Errors

Boundary-location error in extension measurements represents the systematic failure of the probe to signal the proper location of the material boundary of the object and is of the form:

$$\delta_e \;=\; \mp 2 \cdot \delta P_{bx}, \tag{78}$$

where the minus sign is for outside-caliper measurements, the plus sign is for inside-caliper measurements, and δP_{bx} is positive for location of the arbitrary boundary behind the material boundary.

a. *Location of Arbitrary Boundary*: Various phenomena give rise to the direction-dependent boundary-location errors inherent to extension measurements which individually contribute to the aggregate effect and δP_{bx}. In this example, there is the probe radius (that is, the finite extension of the probe tip along the axis of measurement) and the probe pre-travel (that is, the finite displacement of the probe along the axis of measurement after mechanical contact with the object is made, but before the probe transducer crosses a threshold and signals). Assuming that the gross effects of probe radius and pre-travel are eliminated by pre-characterization of the probe, there remain the extension-specific sources of error due, for example, to deformation of the object by mechanical contact and variation of pretravel of the probe with the direction of approach to the object.

Given a three-dimensional touch-trigger probe with a specified pre-travel variation over 360° in the X-Y plane of ± 0.5 μm, the extension-specific error is taken to be:

$$\delta e_{bx} \;=\; 2 \cdot 0.5 \ \mu m \;=\; 1.0 \ \mu m. \tag{79}$$

b. *Compensated Boundary Location*: The contribution to extension-specific error due to location of an arbitrary boundary can in some situations be compensated by an empirical procedure while in the majority of cases a theoretical model of the probe-object interaction is required. In the example being followed here, the probe is assumed to have a residual direction-dependent variation in pretravel of:

$$\delta e_{bt} \;=\; 0.1 \ \mu m. \tag{80}$$

3. Average-Material-Boundary Location

Extension-specific error due to average-boundary location is associated with local variation in the position of the material boundary of the object such that it has no single extension.

In the example being followed here, the steel block is assumed to have a surface-finish peak-to-valley of 0.1 μm (0.067 μm RMS) which when mechanically probed leads to a contribution to extension-specific error:

$$\delta e_{bt} = 0.2 \text{ μm}. \tag{81}$$

4. Sum of Extension-Specific Errors δe

In this example, the total extension-specific error δD is the sum of the errors associated with mislocation of a detected boundary relative to a material boundary and mislocation of a material boundary relative to some simple geometrical shape used to describe the object.

$$\delta e = \Sigma \, \delta e_i = \delta e_{bt} + \delta e_{bm} \tag{82}$$

$$= 0.2 \text{ μm} + 0.1 \text{ μm} = 0.3 \text{ μm}$$

Note that extension-specific error is a purely additive error, independent of the magnitude of the extension.

5. Total Extension Error Δe(δe,Δd)

Combining the sum of the extension-specific errors δe with the total distance error Δd yields the total extension error:

$$\Delta e = \delta e + \Delta d = 1.182 \text{ μm} + 2.77 \text{ ppm} \cdot d \tag{83}$$

For the 100mm-long object, the total extension error, computed and rounded to a significant figure, is:

$$\Delta e_{100} = 1.459 \text{ μm}$$

$$\rightarrow 1.5 \text{ μm} = 15 \text{ ppm @ 100 mm}. \tag{84}$$

E. Summary and Analysis of Errors

This error-budget analysis -- carried out for the case of a commercial-quality displacement-interferometer-equipped coordinate measuring machine used to measure the length of a polished, parallel-faced, 100mm-long, steel block -- illustrates issues specific to the example as well as ones associated with precision dimensional measurements in general. The specific numerical results may be summarized and then analyzed in terms of their relative type-hierarchical, additive-multiplicative, and axiom-specific behaviors.

1. Summary of Numerical Results of Analysis

Table 7 below summarizes results specific to the hypothetical example being examined. Tabulated are errors specific to the type and total errors for each type, on both an incremental ($\pm \delta \ell$) and a fractional ($\pm \delta \ell/\ell$) basis. For comparison, also shown (in italics) is the error in the vacuum wavelength of the laser which provides the reference to the SI unit of length.

Table 7. Summary of Total Errors by Dimensional Type in the Example of the Laser-Interferometer-Based CMM Measurement of Extension of a 100mm Block

Dimensional Type	Type-Specific Error		Total-Error in Type	
	Add $\delta \ell$	Mult $\delta \ell/\ell$	Add $\Delta \ell$	Mult $\Delta \ell/\ell$
Extension	0.3 μm	------	1.462 μm	14.62 ppm
Distance	0.35 μm	3.5 ppm	1.102 μm	11.02 ppm
Position	0.75 μm	7.5 ppm	0.812 μm	8.12 ppm
Displacement	0.06 μm	0.6 ppm	0.062 μm	0.62 ppm
Wavelength	------	------	*0.002 μm*	*0.02 ppm*

2. Analysis of Results

The particular results of the CMM/extension-measurement example illustrate a number of more general patterns in dimensional measurements including: a hierarchy of accuracies, dominance by additive errors and multiplicative errors at the short end and long end of the range respectively, and the minimal contribution of laser-wavelength error to total error.

a. *Hierarchy of Accuracies Among the Types*: The last column, which gives the total error $\Delta \ell/\ell$ by type, illustrates the inherent loss of accuracy as one moves through the progression of types of dimensional measurements. Note that the accuracy of the displacement measurements is down by an order of magnitude from that of the reference wavelength. The accuracy of position is down by another order from there, while the accuracy of distance and extension down another factor of two or three further from there. These trends in these relationships are general. On the extremes, displacement measurements always have the best and more readily obtained accuracy and position and distance are intermediate, while extension measurements always have the lowest accuracy and the most dearly bought.

b. *The Additive-Multiplicative Errors*: The second and third columns, which give the type-specific errors, $\delta\ell$ and $\delta\ell/\ell$ respectively, illustrate the additive-vs-multiplicative character of errors in the different types. On the extremes, the extension-specific errors are purely additive, which means they dominate at short lengths. In contrast, the displacement-specific errors are almost purely multiplicative, which means they dominate at long ranges.

For the hypothetical CMM of the example, these trends are further illustrated by Table 8 which shows errors in ranges of 10 and 1000 mm. Note especially in the table the near-pure additive 1.2 μm (121 ppm) error in extension at 10 mm compared to the near-pure multiplicative 0.26 ppm (0.26 μm) error in displacement at 1000 mm.

Table 8. **Total Error by Dimensional Type at Ranges of 10, 100 and 1000 mm in the Example of the Laser-Interferometer-Based CMM Measurement of Extension of a Block**

Dimension	Error @ 10 mm	Error @ 100 mm	Error @ 1000 mm
Extension	1.2 μm /121 ppm	1.5 μm / 15 ppm	3.9 μm / 4 ppm
Distance	0.9 μm / 91 ppm	1.2 μm / 11 ppm	3.6 μm / 3.6 ppm
Position	0.08 μm / 8 ppm	0.11 μm / 1 ppm	0.35 μm / 0.35 ppm
Displcmnt	0.04 μm / 4 ppm	0.06 μm / 0.6 ppm	0.26 μm/ 0.26 ppm

2. Axiom-Type Matrix Analysis of Results for the CMM Example

The error-budget for the CMM measurement of the 100mm block serves to illustrate the use of the axiom-type matrix of errors for the analysis of the performance of the overall system of machine, object, and techniques for the achievement of precision dimensions. Table 9 shows in the type-vs-axiom matrix format of Table 5 the results of the CMM-extension error budget summarized in Table 6. Also shown, in italics, are examples of the matrix elements not illustrated by the CMM-extension example.

a. *Errors in the Zero of Each Type*: As indicated in Table 9, each of the types of dimensional measurements in the CMM-block example has associated with it a type-specific error of the zero ($\delta\ell_z$) corresponding to a shift of the effective origin of that type. In the displacement measurement, there are the uncompensated change in the index of refraction of the dead-path, the movement of the effective optical-path location of the interferometer reference due to thermal drift, and the uncertainty in the location of the zero due to the least-count of the interferometer system. In the position measurement, there

Table 9.   **The Axiom-Type Error Matrix for the Example of the Laser-Interferometer-Based CMM Measurement of Extension**

Dimension	Error of Zero	Error of Unit	Error of Scale
Displacement	Dead-Path Thermal-Drift Least-Count	Laser Wavelength Mirror-Path Index Cosine I-vs-L	Polarization Mixing *Grad in MP Index* *Bending of MP Ways*
Position	Probe-Tilt Abbe Probe-Setting	Cosine T-vs-I	*Grad in Table Temp*
Distance	Point-Definition	Cosine O-vs-T Object T-Expansion	*Bending of Object*
Extension	Arbitrary Bndry Matl-Av-Bndry *Backlash*	*Thickness-Dependent* Probe "Penetration"	*Approach to Probe* *Resolution Limit*

are the non-cumulative change in the location of the position origin due to the tilt of the probe relative to the displacement axis (Abbe error) and the uncertainty in its location due to variations in probe-setting. In the distance measurement, there is the uncertainty in objects geometry which limits reduction to centroids, that is, points. Finally, in the extension measurements, there are the arbitrariness of the boundary located relative to the material boundary and the variation of the location of the material boundary about its mean. While not in the example, backlash corresponds to an error of the zero of extension because it only appears with the bi-directional approach to the object.

b. *Errors in the Unit of Each Type:* Similarly, as indicated in Table 9, each of the types has associated with it a type-specific error of the unit ($\delta \ell_u$), that is, error proportional to the first-order of the dimension measured.   In displacement, there are the error in the value of the vacuum wavelength of the laser, in the index of refraction of the path of the moveable mirror, and the angle of misalignment between the axis of the interferometer (I) and the direction of propagation of the incident light (L). In position measurement, there is the angle of misalignment between the axis of translation of the mirror (T) and the axis of the interferometer (I). In distance measurement, there is angle of misalignment between the object (O) and the probe-path, which for a rigid-body system is the same as the axis of translation of the mirror (T). While not discussed in the example, an error in the unit of extension can also arise due to any dependence of the effective "penetration' of the probe (positive or negative) on the thickness of the object.

c. *Errors in the Scale of Each Type:* Finally, as indicated in Table 9, for each dimensional type, there are type-specific nonlinearities corresponding to errors of the scale ($\delta \ell_{s1}$). In the displacement measurement, there is the intermixing of the reference- and measuring-beams due to their imperfect separation by polarization states which leads to a non-linearity in the interpolation between fringes. While not in the example, other conditions lead to errors of the scale. In displacement measurements, these include a gradient in the index of refraction along the interferomter mirror path (MP) as well as any lateral bending of the mechanical ways which support the mirror along that path; in position measurements, a gradient in the temperature of the part of the machine which supports the object to be measured; in distance measurements a bending of the object; and in extension measurements, an approach to the resolution limit of the probe.

• • •

In sum, this analysis of the hypothetical case of the use of a laser-interferometer-based coordinate measuring machine for measurement of the length of a part has shown the nature and types of the errors that can arise in that specific example. What the example illustrates is that the use of the axiom-type matrix of errors provides as a complete, self-consistent scheme for systematically partitioning and analyzing such errors. The approach is, however, more generally useful.

3. Application of Type-Axiom Error Matrix to Manufacturing Operations

The axiom-type matrix just applied to the characterization of the dimension-measurement process applies equally to the characterization of the dimension-generation process, that is, the manufacture-to-design of a dimensionally-specified product.

a. *"Machine Tool" versus "Measuring Machine":* Machine tools for the generation of dimensioned forms and measuring machines for the characterization of dimensioned forms share the elements of frame, carriage, scale and "probe," where the formers's probe is for changing the location of material and the latter's "probe" is for ascertaining where the material is located. This commonality is shared by all such "machine tools" — whether metal-cutting lathes, ion-beam milling machines, or photolithographic step-and-repeat cameras — and all such measuring machines — whether they be coordinate measuring machines, measuring microscopes, or micrometer calipers. As such machine tools and measuring machines must conform to the same axiom-type system to carry out their essential functions.

b. *Example of Extension Errors in Milling*: All the errors indicated in Table 9 for the CMM measurement of a block are also manifest in machine-tool-based manufacture of dimensionally specified products.

Consider briefly, for example, the use of a laser-interferometer-based milling machine for the manufacture of a multiple-finned turbine-blade fan where the part design calls for the machining of equally spaced fins of identical tapering cross-section. In general the machine will be subject to all the sources of error identified in Table 9. Even under the conditions that the machine can generate a cutter path that conforms ideally to the part design (that is, can generate displacement, position and distance perfectly), errors of extension can still occur. Under certain conditions of tool-force and material-stiffness, deflection of the fin by the cutter will vary in proportion to the thickness of the fin at the point of cutting, resulting in error in the material form even for an errorless cutter-path. Thus, while the spacing between fins is perfect (corresponding to no errors of the unit, the zero and the scale in distance), the error in the cross-section of each fin would have a "cooling-tower" shape corresponding to errors in each of the zero, unit and scale of extension and fully analogous to the arbitrary boundary, thickness-dependent probe-penetration and limit-of-resolution errors indicated in Table 9.

VII. CONCLUSION

This chapter on precision dimensional measurements for modern manufacturing has presented a new scheme for the assessment of dimensional error, both in measurement processes and in manufacturing processes. The scheme is based on a matrix of dimensional types and measurement axioms which forms a complete, self-consistent system for assessing dimensional error. The scheme has been illustrated by a case-study of a coordinate measuring measurement of an extended object, with the type-axiom related errors indicated as being similarly manifest in manufacturing processes such as the milling of turbine blades. This type-axiom matrix should provide a useful means for the exposure of error, the diagnosis of its cause, and the means of practical elimination through more effective design of machines, processes and techniques for the manufacture of products with precisely dimensioned forms.

ACKNOWLEDGMENTS

The author wishes to acknowledge the generous contribution and evaluation of ideas contained in this chapter by Drs. Estler, Gilsinn, Phillips, Simpson, Stone, Teague and Vorburger of the Manufacturing Engineering Laboratory of NIST.

Appendix A. Errors in Dimensional Measurements Due to Angular Misalignments of System Elements

Accuracy of measurement of dimensions of physical objects by means of a coordinate measurement system based on a laser displacement interferometer, such as the prototypical interferometer shown schematically in a plan-view representation here, depends upon precise alignment of the axes of the laser, interferometer, translating mirror, and the object. Errors in measurements of displacement, position, and distance (and/or extension) arise respectively from misalignment of pairs of axes — interferometer and laser, translating mirror and interferometer, and object and translating mirror, as described individually and in combination below.

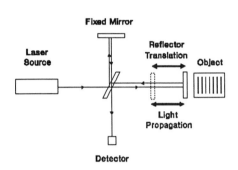

1. Misalignment Error Specific to Displacement

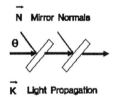

For light of wavelength λ_o propagating in medium of index of refraction n in a direction K at angle θ to the axis of the interferometer (defined by the normals N to the reference and moveable reflectors), the change in spacing of the interferometer mirrors, Δs, is related to the observed change in fringe order number, Δm, by the interferometry equation:

$$D = \Delta s = \Delta m \cdot \lambda_o/2n \cdot \cos\theta . \qquad (A1)$$

Note that misalignment between the axis of the interferometer and the axis of propagation of the incident light constitutes an error of the metric or unit of displacement relative to that of the light.

2. Misalignment Error Specific to Coordinate Position

For a moveable or "measuring" reflecting surface (plane mirror or retroreflector) which translates at an angle ϕ relative to the axis N of the parallel-plate interferometer, the translation T is related to the change in spacing Δs of the interferometer by the "cosine-error" equation:

$$P \cdot \cos\phi = T \cdot \cos\phi = \Delta s . \qquad (A2)$$

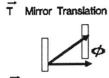

T Mirror Translation

Note that misalignment of the axis of translation of the measuring element of the interferometer relative the axis of the interferometer constitutes an error of the metric or unit of position (coordinate) relative to that of the displacement.

N Mirror Normals

3. Misalignment Error Specific to Object-Distance

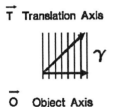

T Translation Axis

Finally, given an object with features with an axis **O** oriented at an angle γ relative to the direction of translation of the measuring mirror of the interferometer, the relation of the distance between features on the object and the translation of the mirror is given by the equation:

O Object Axis

$$d = T \cdot \cos\gamma . \qquad (A3)$$

Note that the misalignment of the axis of the object and that of the translation of the measuring mirror constitutes an error in the metric or unit of distance relative to that of position.

4. Combined Misalignment Error in Distance

For a system in which there are all three misorientations described above, the relation of the distance between features on the object and the observed change in the fringe order number is given by:

$$d = (\Delta m \cdot \lambda_o \cdot \cos\gamma) / (2 \cdot n \cdot \cos\theta \cdot \cos\phi). \qquad (A4)$$

5. Explicit Form for Misalignment Error

Given the sign convention that the observed distance d_{obs} is the algebraic sum of the true distance d_{true} plus the error term Δd, then:

$$\Delta d = d_{obs} - d_{true}$$

$$= d - d(\theta,\phi,\gamma) \qquad (A5)$$

$$= (\Delta m \cdot \lambda_o/2n) \cdot [1 - \{\cos\gamma/(\cos\theta \cdot \cos\phi)\}]$$

Under the conditions that the angles θ, ϕ, and γ are each small, successive substitution of the approximations $\cos x \approx 1 - x^2/2$ and $1/(1 - x^2) \approx 1 + x^2$ into Eq. A5 leads to:

$$\Delta d \approx (\Delta m \cdot \lambda_o/2n) \cdot [1 - \{(1 - \gamma^2/2)/(1 - \theta^2/2) \cdot (1 - \phi^2/2)\}] \qquad (A6)$$

$$\approx (\Delta m \cdot \lambda_o/2n) \cdot [\, 1 - \{(1 - \gamma^2/2)\cdot(1 + \theta^2/2)\cdot(1 + \phi^2/2)\}]\,,$$

which with multiplication and retention of terms of the lowest order yields the final result:

$$\Delta d \approx (\Delta m \cdot \lambda_o/2n) \cdot (\, + \theta^2 + \phi^2 - \gamma^2)/2 \qquad (A7)$$

Note that since the first and second terms are negative and the third term is positive, the overall error in distance due to angular misalignments can be either positive or negative and the measured distance can be either greater than or less than the actual.

REFERENCES

1. E.C.Teague, "Nanometrology", Proc. of Engineering Foundations Conference on Scanning Probe Microscopes: STM and Beyond", Santa Barbara, Jan. 1991, published by The American Physical Society.

2. "Documents Concerning the New Definition of the Metre," Metrologia 19, C. Springer-Verlag, 163-177 (1984).

3. R.C.Quenelle and L.J.Wuerz, "A New Microcomputer-Controlled Laser Dimensional Measurement and Analysis System," Hewlett-Packard Journal, 2-22 (1983).

4. J.S.Beers and K.B.Lee, "Interferometric Measurement of Length Scales at the National Bureau of Standards," Precision Engineering Journal, Vol. 4, No. 4, 205-214 (1982).

5. T.D.Doiron, "High Precision Gaging With Computer Vision Systems," Elsevier, Industrial Metrology 1, 43-54 (1990).

6. R.I.Scace, "Foreign Trip Report to Japan," Center for Electronics and Electrical Engineering, NIST, Gaithersburg, MD, Report No. 720-46 (1989).

7. E.C.Teague, "The NIST Molecular Measuring Machine Project: Metrology and Precision Engineering Design," Journal of Vacuum Science and Technology B7, 1890-1902 (1989).

8. C.F.Vezzetti, R.N.Varner, and J.E.Potzick, "Bright-Chromium Linewidth Standard, SRM 476, for Calibration of Optical Microscope Linewidth Measuring Systems," NIST Special Publication, 260-114 (1991).

9. H.P.Layer and W.T.Estler, "Traceability of Laser Interferometric Length Measurements," NBS Technical Note 1248 (1988).

10. C.Eisenhart, "Realistic Evaluation of the Precision and Accuracy of Instrument Calibration Systems", Journal of Research of the National Institute of Standards and Technology, Vol.67C, No.2, April-June 1963.

11. R.Carnap, "Philosophical Foundations of Physics," An Introduction to the Philosophy of Science, ed., Martin Gadner, Basic Books, Inc., New York, London (1966).

12. J.A.Simpson, "Foundations of Metrology," Journal of Research of National Institute of Standards and Technology, Vol. 86, No. 3 (1981).

13. C.R.Steinmetz, "Sub-Micron Displacement Measurement Repeatability on Precision Machine Tools With Laser Interferometry," SPIE Conference on Optical Mechanical and Electro-Optical Design of Industrial Systems, Vol. 959, 178-0192 (1988).

14. F.E.Jones, "The Refractivity of Air," Journal of Research of the National Institute of Standards and Technology, Vol. 86, No. 1 (1981).

15. W.T.Estler, "Laser Interferometry in Length Measurement," NIST Special Publication, (1990).

16. W.T.Estler, "High-accuracy Displacement Interferometry in Air," Applied Optics, Vol. 24, No. 6, 808-815 (1985).

17. C.R.Steinmetz, "Sub-Micron Displacement Measurement Repeatability on Precision Machine Tools With Laser Interferometry," SPIE Conference on Optical Mechanical and Electro-Optical Design of Industrial Systems, Vol. 959, 178-0192 (1988).

18. G.Zhang, R.Veale, T.Charlton, B.Borchardt, and R.Hocken, "Error Compensation of Coordinate Measuring Machines," Annals of the CIRP, Vol. 34, 445-448 (1985).

19. N.A.Duffie and S.J.Malmberg, "Error Diagnosis and Compensation Using Kinematic Models and Position Error Data," Annals of the CIRP, Vol. 36, 355-358 (1987).

20. "Methods for the Performance Evaluation of Coordinate Measuring Machines", American National Standard B89.1.12 (1991), American Society of Mechanical Engineers, NY, NY 10017.

21. M.T.Postek, "Scanning Electron Microscope-based Metrological Electron Microscope System and New Prototype Scanning Electron Microscope Magnification Standard," Scanning Microscopy, Vol. 3, No. 4, 1087-1099 (1989).

22. R. Donaldson, "Error Budgets", in Tech. of Machine Tools, Vol.5, Ser/ 9.14 (1980).

QUALITY IN AUTOMATED MANUFACTURING

DONALD S. BLOMQUIST

Automated Production Technology Division
Manufacturing Engineering Laboratory
National Institute of Standards and Technology
Gaithersburg, Maryland 20899

I. INTRODUCTION

In this chapter the Quality in Automation (QIA) program is described at the Manufacturing Engineering Laboratory of the National Institute of Standards and Technology (NIST). The purpose of the QIA program is to develop a quality control and quality assurance system that exploits deterministic metrology principles in an automated manufacturing environment, producing small batches with commercially available and affordable equipment. This "deterministic manufacturing" is based on the premise that most errors occurring in the manufacturing process are repeatable, thus predictable. With predictability, the errors can consequently be compensated and quality can be assured through control of both the manufacturing process and the equipment used in this process. The QIA program combines statistical process control methods with on-machine sensing and gauging, real-time error compensation and distributed processing to produce parts of consistently high quality.

The need for a QIA program arises from the limitations involved in traditional manufacturing. Traditional techniques assumed one person running one machine and creating one part over and over again, possibly for years. The operator not only made the part, but was responsible for successful coordination of the entire procedure. This responsibility included any required changes to the manufacturing process prompted by tool wear, machine condition, and material composition.

The operator was also required to adapt to any changes in dimension and tolerance dictated by design. These changes would then be checked by the quality control people using simple micrometers and other single-purpose gauges. Parts that did not meet requirements were rejected, causing a loss in time and money.

To remain competitive in today's global economy, such traditional procedures are insufficient, and indeed changes have already occurred. The operator is now frequently in charge of several machines arranged in what is called a "cell." Design engineers now employ CAD systems rather than drawing boards, and parts are designed and manufactured to finer and finer specifications. Quality inspectors use coordinate measuring machines or CMMs to inspect the finished part. Finally, the machines themselves are controlled by computers using numerical control (NC) code to dictate the steps in the machining process. All these changes allow for greater speed and flexibility in the production of parts, thus reducing waste and driving down costs.

II. OVERVIEW

All of the aforementioned changes create a new environment in manufacturing, changes which have been successfully exploited in large batch manufacturing. However, these changes need to be adapted to the small batch manufacturing environment; this is the purpose of the Quality in Automation program.

Deterministic manufacturing methods rely on in-process measurements that characterize the machining process and form this characterization or model so that the model can be used subsequently to alter and compensate for errors as they occur rather than through post-process measurements on the finished product. All the latter can do is reject parts that are substandard — a clear source of waste in the manufacturing process. Consequently, with deterministic manufacturing methods a responsive, self -adjusting machining system is both a desirable and now possible goal.

There are three necessary aspects to achieving quality: dimension, shape, and surface finish. The Quality in Automation program requires control of the two main components of the manufacturing process: control of the equipment, that is, the machines, and control of the parameters involved in the machining process. Control of the equipment primarily requires the pre-process characterization of the machine tool involved, because, unlike large-batch manufacturers, small-batch manufacturing cannot afford to scrap or remachine parts and so cannot "tweak" the machine after a trial run. All idiosyncracies inherent in the machine must be discovered and accounted for before manufacturing begins. This is possible because machines are by nature more repeatable than they are accurate — thus their "baseline" readings tend to be more stable for thermal characteristics than for geometry.

The second part of the QIA program is to control the other parameters involved in the machining process, recognizing that the process itself is dynamic, with the tool, the part, and the machine all affecting the outcome in quality. These parameters, including tool wear, vibrations, surface finish, forces generated by the machining process, and dimensional integrity must be monitored and controlled.

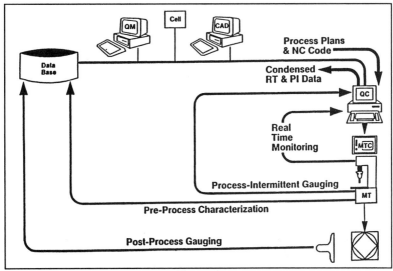

Figure 1

The foundation of the QIA program is the quality control architecture. This architecture uses multiple feed-back loops to control the process as shown in Figure 1. There is a real-time loop, a process-intermittent loop and a post-process loop. This project was devoted to the construction of an overall strategy and the control architecture that would combine traditional part inspection and statistical process control (SPC) methods with the above-mentioned tools of automation (e.g., on-machine sensing and gauging). It must be remembered that research in the QIA project is ongoing. Consequently, although we cannot present the program as finished, we can describe and summarize with confidence the current stage of research as "cutting edge."

The function of the **real-time** control loop is to monitor the machine tool and the metal-cutting process and to modify the tool path, feed rate, and the spindle speed during cutting in order to achieve higher accuracy and surface quality of the workpiece. Monitoring can be done using various sensors incorporated into the machine tool such as position, temperature, vibration, audio and ultrasonic sensors. Real-time error compensation is achieved by implementing tool-path modification of the machine using a combination of kinematic and geometric-thermal (G-T) models of the machine-tool errors. The G-T model is constructed from the **pre-process machine characterization** measurements. The kinematic model is constructed, based on the machine structure, using the theory of rigid body kinematics to describe the relative relationships between machine elements. The required correction to compensate for the resultant error, which is calculated using these models, is implemented by the Real-Time Error Corrector (RTEC). The RTEC is a microcomputer-controlled device which is inserted between the position feedback device for each axis and the machine tool controller. It alters the feedback signal to cause the machine to go to a slightly different position to compensate for the predicted errors. The feed-rate and the spindle-speed modifications will be made either to minimize the vibration and chatter during cutting, or to optimize surface finish.

The functions of the **process-intermittent** control loop are 1) to determine workpiece errors introduced by the machining process which cannot be compensated by the real-time control loop and 2) to correct for them by generating modified NC part program and/or tool dimension offsets for the finishing cut. On-machine dimensional and shape measurements of the workpiece are performed (between semifinishing and finishing cuts) using fast probing. Fast probing is a gauging method which uses a touch-trigger probe at feed rates 10 to 20 times higher than is currently used. This is important because it takes less time to take data points for M-machine inspection. The functions of the **post-process** control loop are to verify the cutting process and to tune the two other control loops by detecting residual systematic errors in the process and generating corrective actions. Process verification, over a period of time, is done by inspecting the features of finished parts independently using a coordinate measuring machine. The errors measured on similar features, e.g., circles, are then correlated back to machine tool errors and the process parameters. Systematic residual errors are used to determine modifications to the algorithm used in the process-intermittent control loop to alter the finishing cut based on the errors detected by on-machine gauging. In the long term, systematic errors of geometric features measured in the post-process control loop will be used to modify the geometric-thermal model used in the real-time control loop or to indicate that the pre-process characterization of the machine tool geometric errors needs to be revised.

At each machine tool, the Quality Controller (QC), a PC-compatible computer, is interfaced to the machine tool controller through RTEC. Using the geometric-thermal model and a kinematic error model of the machine structure, the QC calculates the resultant error vector. This is converted to corrections required for each axis in units of correction counts and sent to the RTEC. The RTEC implements these corrections and reports the current correction status back to the QC. In the process-intermittent control loop, the QC receives the trip-point axes positions from the RTEC and stores this data. When probing is complete, the QC determines the shape and the dimensions of the gauged part and calculates the modification to the tool offsets or NC-part program coordinates to be used for the finishing cut. For the post-process control loop, the PC at the CMM is used off line to generate the inspection program in a Dimensional Measuring Interface Specification (DMIS) format. The inspection results are reported back to the PC in DMIS format for analysis.

The following sections of this chapter describe in detail the four major aspects involved in research in the QIA project: kinematics, the G-T model, the RTEC, and the three control loops in the QIA architecture. The development of the kinematic model derived from the theory of rigid-body kinematics is described in some detail, laying down the theoretical basis. From this the Geometric-Thermal (G-T) model for the specific machine was derived. This paved the way for use of the Real-Time Error Corrector. The RTEC was then integrated into the real-time control loop of the QIA architecture.

After completion of the final integration of the real-time and the process-intermittent control loops for the turning center, efforts are focused on the post-process control loop of the QIA architecture. The specific tasks of this effort are development of a feature-based Quality Database, development of a Quality Monitor, and integration of the Quality Database with the Quality Monitor. The Quality Database will be used to store all the information about the parts produced in the system based on their specific features. Such information will consist of: the errors found in these features, the machine axes, cutting tools, and cutting parameters used in producing these features, as well as other sensory information obtained during cutting. Examples of sensory information include temperature profiles, vibration signatures, and on-machine probing data. The Quality Monitor will be designed to sort this information stored in the Quality Database to identify the residual systematic errors in the process and generate corrective actions.

In the following sections, we describe the machine characterization effort, which is the combination of the development of kinematic model and the analysis of the raw geometric and thermal data to build the G-T model. We describe the tests designed and carried out for the evaluation of the RTEC's real-time tool-path modification performance. Finally, we describe the process-intermittent loop as well as the post-process loop which involves coordinate measuring machines.

III. PRE-PROCESS MACHINE CHARACTERIZATION

The purpose of characterizing a machine tool is to determine the actual tool or cutter position relative to the workpiece under transient thermal conditions. These positioning errors establish the machine tool accuracy and the precision of the workpiece dimensions. The total positioning errors of the tool relative to the workpiece are the result of a complex interaction among the machine tool elements, the machining operations, and the dynamic environment within which the machine tool operates. Due to this complexity it is necessary to determine the components of the total positioning errors and the dependence of these components upon the machining operations and the resulting environmental conditions. With this knowledge, it is possible to characterize the machine tool and thus improve machining accuracy [1-10].

A. The Kinematic Model

Developing a mathematical description of position errors constitutes the first step in the QIA program. A model is constructed, representing the positioning error of the cutting tool with respect to the workpiece as a combination of the individual machine component errors.

Machine tools are multidegree of freedom structures which consist of stacked slides and spindles. Each moving element of the machine tool has error motions in six degrees of freedom, three translations and three rotations. Therefore, depending on the number of axes (slides), a machine tool can have as many 30 individual error components. Furthermore, the effect of the error of one slide may be amplified by the motion of another slide. On the other hand, the end effect of all these individual error components is a single translation and a rotation of the cutting tool around an arbitrary axis in space. Thus, for a single-point cutting operation, a simple translation is enough to compensate for the errors toward achieving higher machining accuracy.

With the increased computing power of microprocessors, total software-based error compensation for machine tool errors (based on the mathematical error models) becomes feasible. The mathematical error models created for this purpose should be easily transformed into a modular and maintainable software. Furthermore, these error models should be in a general format so as to be easily adapted to different machine tool configurations. Based on these criteria, a generalized approach was developed for the total error model for a machine tool, which systematically combines all the error components and yields an error vector. In developing this model, with the assumption of rigid body kinematics, a machine tool-fixture-workpiece system is considered as a chain of linkages and spatial relationships are described between these linkages using homogeneous coordinate transformation matrices. This model can be applied to any machine tool or any type of positioning machines such as coordinate measuring machines or robots [11].

In general, a homogeneous coordinate representation is defined as the representation of an N-dimensional position vector by an (N+1)-dimensional vector [12]. The (N+1)st component of this vector is called a scale factor. In a homogeneous coordinate representation, the actual components of an N-dimensional vector can be found by dividing each component of the (N+1)-dimensional vector by the scale factor. In a 3-dimensional space, the homogeneous coordinate representation of a vector $p = (p_x, p_y, p_z)^t$ is $p' = (p_x, p_y, p_z, 1)^t$, of which the scale factor is selected as unity.

A homogeneous coordinate transformation matrix in 3-dimensional space is a 4x4 matrix. It is used to express a homogeneous coordinate vector in one coordinate system with respect to another coordinate system. Similarly, a homogeneous transformation matrix, T, can be used to represent one coordinate system with respect to another or reference coordinate system.

With this matrix, it is possible to describe the relative rotation and translation between any two coordinate frames. If the coordinate frame is embedded in an object, then the matrix, T, describes the relative position and orientation of this object with respect to another object or coordinate frame in space. An important feature of homogeneous coordinate transformations is that they can be multiplied in series to describe one object with respect to several different coordinate frames. This feature is very useful in describing structures which consist of several elements (linkages) that are positioned with respect to each other [13-16].

1. The Model For Machine Tools

Since a machine tool can be considered as a chain of linkages, an approach of describing the spatial geometry of linkages with respect to a reference frame by matrix multiplications can be used to determine the spatial relationship between the cutting tool and the workpiece. As shown in the previous section, any element of the machine tool can be represented by a homogeneous transformation matrix, which describes the relative position and orientation of a coordinate frame embedded in this element with respect to another coordinate frame embedded in another element of the machine. Thus, by multiplying the homogeneous transformation matrices corresponding to a series of elements such as carriage, cross slide, cutting tool, workpiece and spindle, one can describe the cutting tool and the point on the workpiece which has to be in contact with the cutting tool, with respect to a conveniently chosen fixed reference frame. Since the cutting tool follows the contours of the ideal workpiece geometry, the resultant homogeneous transformation matrices, T_{TOOL} for the cutting tool and T_{WORK} for the workpiece, should be identical.

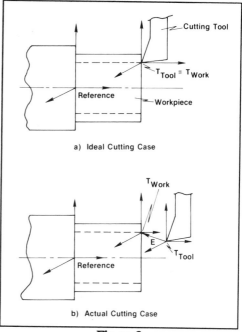

Figure 2

However, due to the errors involved in the machining operation, the two matrices are not identical (See Figure 2). The total error, E_{total}, which causes an inaccurate cutting operation, is represented by the following equation:

$$T_{WORK} = T_{TOOL}\, E_{total} \tag{1}$$

Thus, the equation

$$E_{total} = T^{-1}_{TOOL}\, T_{WORK} \tag{2}$$

gives the resultant error matrix. Application of this approach to a two-axis machine tool is given in the following. Consider a two-axis turning center. The schematic of this type of machine tool is shown in Figure 3. The structure of the machine consists of a spindle, which is connected to the bed of the machine with a revolute joint, a carriage, connected to the bed with a prismatic joint, a cross slide, which is connected to the carriage with a prismatic joint, and a tool turret, which is connected to the cross slide with a revolute joint. Note that motion on the tool turret joint is restricted when there is a cutting action. Finally, there are also a cutting tool, which is rigidly connected to the tool turret, and a workpiece, which is rigidly connected to the spindle. Figure 3 also shows the locations of the coordinate frames, which are assigned to each element of the machine-tool-workpiece system.

Using transformation matrices, the cutting tool with respect to the reference frame is represented by the following matrix multiplication:

$$T_{TOOL} = {}^0T_1 \, {}^1T_2 \, {}^2T_3 \, {}^3T_4 \tag{3}$$

Similarly, the ideal workpiece-cutting tool interface point on the workpiece with respect to the reference frame is given by

$$T_{WORK} = {}^0T_5 \, {}^5T_6 \tag{4}$$

Finally, the error matrix **E** is calculated using Equation 2. The fourth column of this matrix represents the position of the tool with respect to the workpiece, which is expressed by its three orthogonal components, $p_{ex}, p_{ey},$ and p_{ez}. Since a two-axis turning center does not have any ability to move the cutting tool in the third orthogonal direction; i.e., the "y-axis," the third component of the resultant vector is of no significance to the current error compensation effort. These errors in the "insensitive" direction do not need to be considered. Therefore, only two components of the error vector are given by the following equations:

$$p_{Ex} = \epsilon_y(s)*z(w) - [\epsilon_y(z) + \epsilon_y(x)]*Z_T - \delta_x(z) - \delta'_x(z) - \alpha_p*\Delta z + X_1 \tag{5}$$

$$p_{Ez} = -\epsilon_y(s)*x(w) + [\epsilon_y(z) + \epsilon_y(x)]*X_T - \epsilon_y(z)*x - \delta_z(z) - \delta'_z(x) - \alpha_o*\Delta x + Z_1 \tag{6}$$

where

$\epsilon_y(s)$	tilt error of spindle about y-axis,
$\epsilon_y(z)$	yaw error due to carriage z-motion,
$\epsilon_y(x)$	yaw error due to cross slide x-motion,
$\delta_x(x)$	displacement error of cross slide x-motion,
$\delta'_x(z)$	x straightness of carriage z-motion,
α_p	parallelism error between z-motion and axis average line of spindle,
$\Delta x, \Delta z$	incremental x and z motion,
$\delta_z(z)$	displacement error of carriage z-motion,
$\delta'_z(x)$	z straightness of cross slide x-motion,
α_o	orthogonality error between x-motion and axis average line of the spindle,
$x(w), z(w)$	ideal cutting point coordinates on work piece,
X_1, Z_1	machine offset,
X_T, Z_T	tool dimensions.

In order to determine the resultant error vector of the machine tool, the individual error components in the above equation have to be identified. These error components are functions of the values of x and z, and of the machine temperatures as defined by the empirical G-T model described in the following section. Once the error vector is computed, it can be used for error compensation purposes. Other than error compensation purposes, this model can be used to evaluate the performance of the machine tool, or it can be used during the design stage of the machine tool to evaluate the effects of different geometries on the overall accuracy of the machine tool.

B. The Geometric-Thermal (G-T) Model

Among the major factors affecting the positioning errors are the geometric errors of the machine tool and thermal effects on the geometric errors [17-24]. Geometric errors are caused by unwanted motions of the machine such as carriages, cross-slides, and worktables. These unwanted motions occur because of geometric imperfections and misalignments. Heat generated by the machine tool and the cutting operation causes temperature changes of the machine tool elements and environment. Due to the complex geometry of the

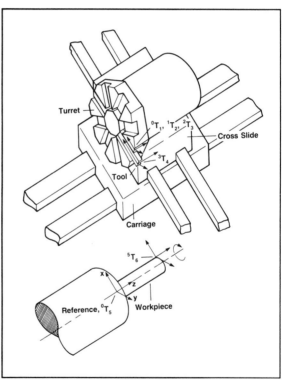

Figure 3

machine structure, concentrated heat sources (e.g., drive motors and spindle bearings) create thermal gradients along the machine structure. Spindle growth, lead-screw expansion, and a significant part of the machine structure deformations are the results of these temperature changes and gradients. Therefore, a G-T model is a collection of mathematical functions which describe the characteristics of the machine's geometric errors under different thermal conditions.

To characterize the turning center's quasistatic and thermally induced geometric errors, a statistical data analysis procedure was developed. This procedure was used to establish error functions with respect to machine positions and the temperature profile. The premise is that the error functions are continuous, and the error values are slowly changing with respect to positions and temperatures. In addition, from a previous investigation it was found that the cause and effect relationship between structure temperature and the resultant error is relatively simple [25].

1. Data Acquisition and Machine Tool Metrology

In order to predict the resultant error at any location and at any time on the machine work zone, all of these error components must be predicted [26-32]. Although these components are geometric error components of the structural elements of the machine tool, their characteristics change as a result of thermal effects, loading conditions, etc [33-40].

Since two independent variables, the nominal position and the thermal state of the machine, affect the geometric errors, measurements are carried out to find the relationships between these variables over their possible range. For each machine axis, the measurement starts when the slide under study is at one end of its travel range. Then the slide moves toward the other end of its travel range while a reading is taken at every measuring interval. The motion is reversed at the end of the travel, and the slide is sent back to its starting position again, with a reading taken at every measuring interval. With this procedure, it is possible to measure the reversal error on each axis. This measurement is necessary (even though leadscrew backlash compensation is achieved in some commercially available machines) due to the fact that this type of backlash compensation considers neither the nonlinearities of the leadscrew nor the thermal effects. In selecting the measuring intervals, one has to consider the periodic error components such as those caused by the leadscrew misalignment. In order to eliminate the effect of the periodic errors, the measuring interval should be selected at even multiples of the leadscrew lead. However, in order to determine the periodic error measurements they should be taken separately over a very short travel range with the assumption of uniformity over the whole travel range.

The error components are classified into four groups with respect to characteristics, measurements procedures, and the sensors used. These are linear displacement errors, angular errors, errors in straightness, parallelism and orthogonality, and errors resulting from spindle thermal drift.

The linear displacement errors are measured using laser interferometry, angular errors are measured with electronic levels to measure the roll error of machine slides, with the remaining angular errors like pitch and yaw measured best by laser interferometry. Straightness-related errors may be measured with high precision proximity probes as well as laser interferometry, while spindle-related errors require high precision proximity probes.

2. Data Analysis

The following section describes the data analysis carried out for a two-axis turning center. Similar analyses can be conducted for other types of machine tools. The machine-characterization raw data consists of a series of error values taken at each measuring interval while a particular machine slide was moved back and forth along its axis of motion. The temperatures at as many

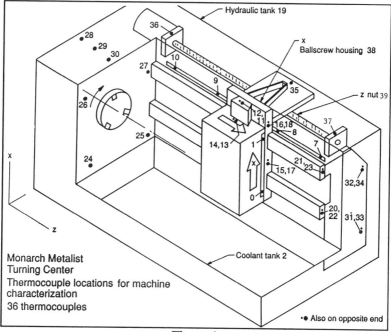

Figure 4

as 36 locations around the machine structure were taken before and after each bidirectional run. Figure 4 shows the temperature measurement locations on the machine. The descriptions of these locations are given in Table 1. The procedure for data acquisition is as follows. Since the machine is programmed to move in predetermined discrete steps along its travel axis, direct communication between the data acquisition computer and the machine tool controller is not necessary. The data acquisition computer should, however, know when to read the laser interferometer during the geometric error measurement cycle. We established the needed synchronization by activating "M" codes at the end of each motion step in the NC program. "M" codes are part of NC program syntax which are used to execute special functions in the machine controllers. In response to the activation of the synchronization M codes, relays switch a corresponding signal level, which can then be monitored. Through this monitoring, the data acquisition computer determines when to read the laser interferometer data. A sample plot of the raw data obtained from the z displacement error measurements is shown in Figure 5. The apparent nonrepeatability of the data is due to the changes in the temperature profile of the machine.

We started the data analysis by normalizing the error values at each nominal position with respect to the measured error at the starting position of each bidirectional run. Figure 4 shows the normalized version of the same data shown in Figure 5. Since the error data is direction-sensitive due to backlash and other reversal errors, we separated the data into two groups according to direction.

For each nominal position, we had one set of error data for the forward direction and one set for the reverse direction. By interpolating the

the z displacement error of the turning center.

The error surfaces generated by the statistical techniques described above can be used to predict any particular geometric error component of the machine during the cutting operation regardless of the time and/or temperature history of the machine.

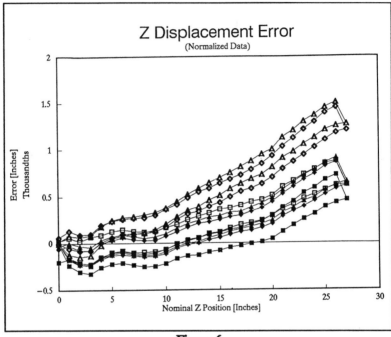

Figure 6

TABLE 1

Thermocouple Locations on Monarch Metalist Turning Center

NO.	LOCATION
0	Bottom of X-(Glass) Scale
1	Top of X-Scale
2	Coolant Tank
3	Ambient
4	Not Used
5	Not Used
6	Top Right of Bed
7	Right of Z-Scale
8	Right Center of Z-Scale
9	Left Center of Z-Scale
10	Left of Z-Scale
11	Top of X-Way
12	Bottom of X-Way
13	Top of X-Head
14	Bottom of X-Head
15	Bottom of Z-Slide
16	Top Left of Z-Slide
17	Bottom Right of Z-Slide
18	Top Right of Z-Slide
19	Hydraulic Tank
20	Left End of Lower Z-Way
21	Left End of Upper Z-Way
22	Right End of Lower Z-Way
23	Right End of Upper Z-Way
24	Lower Front of Spindle Head
25	Lower Rear of Spindle Head
26	Upper Front of Spindle Head
27	Upper Rear of Spindle Head
28	Left of Top of Spindle Head
29	Middle of Top of Spindle Head
30	Right of Top of Spindle Head
31	Bottom Left of Bed
32	Top Left of Bed
33	Bottom Right of Bed
34	Not Used
35	Near X-Drive Motor Shaft Bearing
36	Left Z-Ballscrew Bearing
37	Right Z-Ballscrew Bearing
38	X-Ballscrew Housing
39	Z-Ballscrew Nut

TABLE 2

Linear Regression Fits of Z Displacement Error as a Function of Different Temperatures
(at position Z=356 mm; forward direction of motion)

THERMOCOUPLE NO.	STANDARD DEVIATION (m)
12	0.86
15	0.86
14	0.89
17	0.91
22	0.95
33	0.95
38	0.97
11	0.99
18	0.99
16	1.0
35	1.1
1	1.1
0	1.1
20	1.2
32	1.3
31	1.33
25	1.47
29	1.48
30	1.49
28	1.52
6	1.59
26	1.62
23	1.66
19	1.69
3	1.70
24	1.71
13	1.80
27	1.82
8	1.86
9	1.90
21	2.12
7	2.13
10	2.28
2	2.56
39	2.73
36	3.37
37	3.73

Table 3

Results of the Statistical Analysis of Z Displacement Error Data (Forward Direction of Motion)

General form of the equation:

$$e(z,T_{12}) = A + BT_{12} + Cz + Dz^2$$

Regression Parameters		Position Segments		
		25 - 250 mm (1 - 10 in)	225 - 475 mm (9 - 19 in)	450 - 680 mm (18 - 27 in)
d Error of Y Estimate (μm)		1.39	1.26	1.26
Squared		0.77	0.95	0.98
umber of Observations		280	308	280
egrees of Freedom		276	304	276
onstant Term (A) (μm)		-3.53	-8.863	-149.127
oefficients	Temp.(B) (μm/C)	1.212	3.029	4.952
	Position (C) (μm/m)	32.703	73.05	96.356
	(Position)^2 (D) (μm/m^2)	-14.961	-30.58	-19.92
ax Positive Deviation (μm)		3.025	2.937	3.916
ax Negative Deviation (μm)		-4.442	-4.027	-3.797

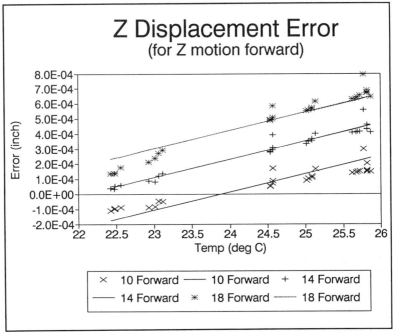

Figure 7

IV. THE REAL-TIME ERROR CORRECTOR

By compensating for systematic errors during the machining process, it is possible to reduce the errors in the finished part. But compensation is not a trivial solution. Corrections can be accomplished in several ways. There can be a hardware modification of the machine tool controller; there can be a software modification of the machine tool controller. A problem arises in that one must know the detailed design of the machine tool controller, often not available due to proprietary concerns. Therefore, a more generic solution was necessary. The Real-Time Error Corrector (RTEC) is that solution.

The function of the RTEC is to allow real-time corrections of positions of the machine tool axes. The objective is to perform this without intrusion into the machine tool controller and in a manner that is invisible to the machine tool. This means that the signals from the position feedback elements are modified to make the machine axes move to slightly different physical positions (as commanded by signals from the QC) than they would without the corrections applied. The RTEC is a device to facilitate error correction operations. The RTEC can be inserted between the position feedback elements (e.g., resolver, encoder, glass scale, etc.) of the axes of a machine tool and the machine tool controller. This device independently and simultaneously counts the signals from the feedback element to produce the machine position. It also includes an interface to send the position data to the QC computer which calculates the required error compensation for each axis. The tool path can be modified by the RTEC which is inserted between a position feed-back device producing "encoder-type" signals and the machine-

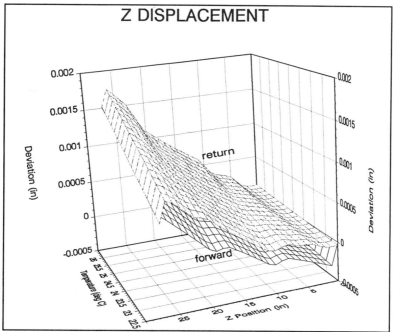

Figure 8

tool controller (MTC). Pulses are added or subtracted from the signals which alters the count in the MTC and thus alters the tool path from the NC-programmed positions. The number of pulses added or subtracted compensates for the resultant error vector computed by the QC from the G-T and kinematic models.

The RTEC is a microcomputer-based device, containing one microcomputer and one position decoder for each axis of the machine tool. Thus, there may typically be two, three, or more microcomputers in the RTEC used for the error correction of one machine tool, with each one communicating in turn with the QC computer as shown in Figure 9. The implementation of the RTEC appears in Figure 10. Quadrature position-feedback signals (square-waves A and B, displaced in phase by 90 deg) from the machine tool axis will pass through the RTEC. A microcomputer (for each axis) will control the operation of the RTEC. This device, an 8-bit microcontroller from the Intel 8051 family, will send similar quadrature signals on to the machine tool controller, and depending upon the mode of operation, will communicate back and forth with the QC.

In addition to its major operation mode of error correction, (described above), the RTEC has two other modes of operation. One is a "transparent" mode. In this mode the RTEC merely passes the quadrature position-feedback signals, unchanged, to the machine tool controller. The transparent mode is required for conventional uncorrected operation of the machine tool, and is necessary when the QC is disconnected from the RTEC. The other mode is the probing mode. In this mode, the probe-trip signal "holds" the trip position at the position decoder output until the microcomputer can read the position and output it to the QC. This occurs while the decoder continues to

count the encoder signals to track the machine axis position. This device
enables the carrying out of fast probing critical to minimizing the on-machine
gauging time during process intermittent inspection.

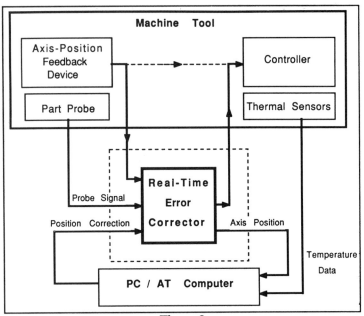

Figure 9

Fast probing is an on-machine part measurement using a touch-trigger
probe with a feed rate of 2500 mm/min (100 in/min) which is 10- to 20-times
faster than those more commonly used.   To determine the measurement
repeatability of fast probing, a point on the face of a part was probed
repeatedly.   The stand-off distance was 5 mm (0.2 in) and the over-travel
distance 2.5 mm (0.1 in) to ensure a constant velocity of 2500 mm/min
(100 in/min) at the trip point.   A point was probed and the trip-point axes-
positions stored in a PC-type computer, interfaced to the RTEC, every
520 ms. The standard deviation of 35 measurements was calculated to be 0.45
micrometer. Hence, the two standard-deviation repeatability is 0.9 micrometer
which is essentially the same as the repeatability specification of the probe
[1 micrometer at a velocity of 480 mm/min (18 in/min)] and the resolution of
the position feedback after decoding (1 micrometer).   Therefore, it can be
concluded that the fast probing does not reduce repeatability.

V. PROCESS-INTERMITTENT ERROR COMPENSATION

In the process-intermittent (PI) control loop, part-program modifications compensate for certain classes of process-related errors by modifying variables that determine the tool-path (such as coordinates, tool offsets, etc.) in an NC part program for the finishing cuts. The differences between the part dimensions gauged during pauses between machining passes, and the corresponding nominal dimensions, are analyzed to determine tool-path adjustments required for subsequent machining passes. The adjustments will be in the form of tool offsets in simple cases; otherwise, the coordinates in the NC part program will be modified. By programming coordinates which were modified on the basis of observed error patterns, the tool will traverse a path closer to the nominal path than it would if the part program remained unchanged.

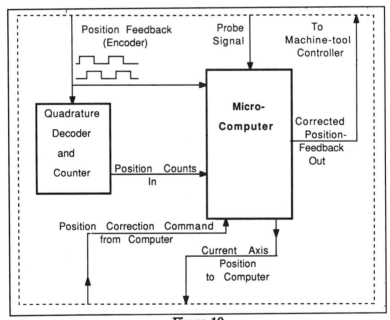

Figure 10

Software for process-intermittent error compensation has been designed as part of an overall strategy which includes real-time error compensation. Explanations of the process-intermittent software will be in the context of a real-time system. One goal of the real-time control strategy of QIA is to compensate for all the systematic machine tool errors using the geometric-thermal (G-T) model and the Real-Time Error Corrector (RTEC). However, some machining errors, such as those due to varying tool length, deflections and wear, will occur in addition to random errors, despite the use of the G-T model and the RTEC. These are the errors targeted by the process-intermittent error-compensation strategy.

A. Methodology for Process-Intermittent Error Compensation

The methodology described in this section has general applicability to turning processes, and nearly all of the principles are extensible to milling. However, the prototype software has been developed for the turning center. The software is designed to accommodate several machine-then-gauge iterations. It is written in the C programming language and runs on the Quality Controller (QC). A high-level programming language environment, AMPLE, is used on the QC to supervise the execution of process-intermittent error-compensation functions. AMPLE was conceived as an interpretive programming language environment which provides off-line programming services to permit the construction of control interfaces to industrial manufacturing systems [41,42].

Process-intermittent dimensional inspection is performed with a touch-trigger probe. The numerically controlled machine tool can be programmed to use such a device to measure part dimensions and profiles, hole diameters, and distances between hole centers. As a gauging routine is executed, the RTEC reports the machine-axis positions to the QC for each probe-trip. The coordinates of the point of contact are then calculated by the QC in terms of the part coordinate system so that they may be compared to the corresponding nominal dimensions.

The QC and the RTEC continuously correct the tool path during machining. But during gauging, the QC switches off this correction function. Therefore, to determine the true probe-trip locations, the gauging data must be adjusted off-line using the G-T model.

The error, or difference between the nominal dimension and the measurement obtained with the probe, is due to differences between the characteristics of the tool and the probe and between the dynamics of the machining and the probing process. Other errors are not detected. The process-intermittent error-compensation strategy described here assumes the G-T model to be accurate. The probe and gauging process are also considered to be accurate for the process-intermittent error data analysis. Inaccuracies in those areas will be detected in the post-process control loop. Because of these idealistic assumptions, the measured error is multiplied by a weighing factor of less than 100% (the actual value to be refined with experience), before being used to adjust the tool path.

B. Implementation of Process-Intermittent Error Compensation

To implement process-intermittent error compensation some critical steps have to be performed in sequence. These steps are segmentation of NC part programs, development of on-machine inspection routines, generation of archival part programs, generation of data files associated with the parts, collection and analysis of on-machine inspection data, and modification of part programs.

1. NC Part-Program Segmentation

In order to insert on-machine inspection routines before and after various semifinishing and finishing cuts, original NC part programs must be divided into segments. Segmentation is based on subdividing the part geometry into features according to the processes for machining different sections. Places in the part program where execution may be safely interrupted must be identified. They typically occur after blocks of code corresponding to the machining of certain sets of features. Since the dimensions after roughing cuts are not of critical interest, NC code pertaining to roughing cuts may comprise large segments, or even a single segment. NC code for semifinishing and finishing cuts, on the other hand, is contained in small segments to permit gauging after only few cuts.

These independent segments are loaded down to the machine-tool controller sequentially, and they are executed one at a time. Therefore, header and footer blocks are added to each segment as appropriate. The end of each segment is marked by an "M30" NC command, which causes the spindle, coolant and feed to stop. Between the segments the execution will be suspended to perform certain logistical tasks for error compensation such as acquiring temperature data, transferring measurement data, and other operations.

2. Development of Dimensional Gauging Routines

In this chapter, the word "cut" refers to: a machining pass of the tool along a single curve (or straight line), or the surface produced by that pass, as a geometric subfeature of the part. In preparation for developing gauging instructions to be sequenced with other segments of the part program, each part surface which will need dimensional measurements must be identified as a cut. Figure 11 shows a sample part with cuts and gauging points identified on it. In this figure, cuts 1 and 7 are "hypothetical cuts." They do not represent actual surfaces on the part, but are defined such that their intersections with real cuts are start- and end-points of tool paths.
Software would assist with the following procedures:

1. On the basis of the part geometry, determine all the measurements that will need to be taken on the surfaces considered. Develop a file of the nominal dimensions at each measurement point.

2. Define the gauging steps for making the necessary measurements. Taking the machine-tool acceleration distance, and overtravel limits of the touch-probe stylus into consideration, the coordinates of the probe paths will be based on the nominal dimensions.

3. After the verification of the probing steps, convert the steps into routines; i.e., entire sequences of formal instructions. Add header and footer blocks to each gauging routine as necessary to form independent part-program segments. Each routine will be executed during a single interval between machining passes.

4. Insert each gauging routine segment into its appropriate place in the sequence of part-program segments.

3. Generation of Archival Programs and Related Data Files

Archival part programs are special adaptations of original NC part-program segments. They are generated by replacing tool-path coordinates with variables and by adding tool-offset commands near the

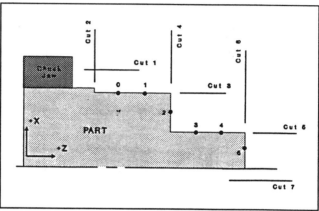

Figure 11

beginning of the program segment. Archival programs function as templates. They are used in conjunction with special data files to generate usable NC part programs. Process-intermittent modifications are made only to temporary copies of the archival part program.

In order to generate the above-mentioned special data files, the following determinations must be made prior to the operation. The x-length of the probe and its stylus diameter must be calibrated. Each part surface to be identified as a cut, as well as the nominal coordinates of the gauging points to be located along the cut, must be decided. The order in which the gauging points will be measured must be known. The point whose z-coordinate is to be used as a reference location on the part must be identified. The tool-path specifications which are to be variable must be decided. The nominal shape of each surface and its important planar relationships to others must be specified.

Among other data available in files is information concerning each cut containing gauging points, including (1) a cut-identification number to identify each surface to be gauged; (2) the particular tool-path specifications responsible for the last machining pass that produced the cut; (3) the direction of the last pass; (4) the surface-normal vector at each gauging point; and (5) codes associating part-program instructions with particular gauging results.

4. Collection and Analysis of Gauging Data

While the machine tool controller is executing a gauging routine, inspection data are acquired by the QC from the RTEC and are stored in a file [43,44,45]. The process-intermittent error compensation software module reads the data from the file and processes it to generate NC-program modifications. The procedure for the analysis is given below.

The error at each measurement point is defined as the vector between the measured coordinates, and the nominal coordinates corresponding to that point. This vector is decomposed into its components along the surface normal and the surface tangent. Only the component along the surface normal is used for compensation purposes. Excessive magnitude of the component along the surface tangent indicates a gauging problem. In this case, a warning should be generated and data corresponding to this point disregarded.

At the end of execution of each process-intermittent gauging routine, the following steps are taken:

1. Using the machine-tool temperatures that existed at the time of gauging, the machine-tool coordinates that are registered at each probe-trip will be adjusted using the G-T model. These adjusted gauging coordinates are then converted from the machine-tool coordinate system to the part coordinate system. The surface normal vector for each gauging point is used for this conversion. Figure 12 illustrates the determination of the part coordinates.

   Referring to Figure 12, X_{part} and Z_{part} can be calculated using the following equations:

   $$X_{part} = X_{mach} + r(1\text{-}\sin\theta) - X_{probe} - \Delta X_{sys} \qquad (7)$$
   $$Z_{part} = Z_{mach} + r(1\text{-}\cos\theta) - Z_{probe} - \Delta Z_{sys} \qquad (8)$$

   The adjusted part coordinates would be identical to the target coordinates in the case of zero error. Otherwise, each error is calculated as the difference between an adjusted coordinate and the target coordinate corresponding to it. For the next machining cut, the tool-path error expected on the basis of this analysis will be compensated by an adjustment in the tool offset or part-program coordinates:

   $$\text{adjustment} = \text{-(error)} * \text{(weighting factor)} \qquad (9)$$

   There may be a need for a separate adjustment equation for each axis, or even for different coordinate ranges for each axis. The equations will differ only in the values of weighting factors, which should be less than 100% in every case, to avoid overcompensation. Coordinates in tool-path specifications in the next machining segment may be adjusted by adding the calculated adjustment to the nominal coordinates in that segment.

2. Modified tool paths should retain the general shape of the original tool paths. Therefore, the set of coordinate adjustments calculated in the above step has to be processed to ensure the smoothness. The adjusted coordinates are fitted to the nominal shape of the measured part feature using least-squares curve fitting techniques. Thus, the actual position and the orientation of the measured feature are determined with respect to those of the ideal feature. NC-program modifications are performed based on this information. The general polynomial equation used to determine the correction curve is given below.

$$C_1z^3 + C_2z^2 + C_3zx + C_4x^2 + C_5z + C_6x + C_7 = 0 \qquad (10)$$

With the use of appropriate coefficients, Equation 10 can be used to describe various geometries such as lines, circles, quadratic and cubic curves. The coefficients are selected based on the nominal shape of the measured feature. The correction curve, derived by fitting the adjusted coordinates to the polynomial, becomes the tool path for the corresponding cut in the next machining segment.

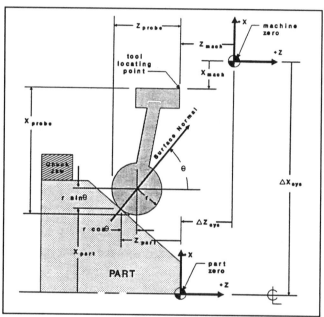

Figure 12

5. Modification of Part Programs

The adjusted data described in the previous section are used as input to correction functions to calculate new tool-path specifications. Figure 13 shows the functional block diagram of the overall part-program modification system. The procedure for the modification is described below.

1. If the measured feature is found to be only translated with respect to the ideal one, which is indicated by a nearly constant offset error along an axis, a tool offset change would be adequate.

2. If the measured feature is found to be translated, rotated, and/or deformed without changing the planar characteristics (called a plane angle error), compensation can be done by modifying selected end coordinates related to these features.

3. If the shape of the measured feature is found to be deformed with respect
    to that of the ideal one, the compensation requires more elaborate
    modifications in the NC programs. These modifications include breaking
    the features into smaller linear sections and modifying the end coordinates
    of these linear sections accordingly.

If tool offsets are to be adjusted, the calculated x and z increments are
substituted into the update commands in the pertinent part-program segment.

For cases in which the endpoint of one smooth tool path also serves as
the start point of the next tool path, the adjusted coordinates for that point
are calculated as the intersection of the two correction curves for the paths.
The new coordinates are substituted for the nominal coordinates in the
pertinent part-program segment.

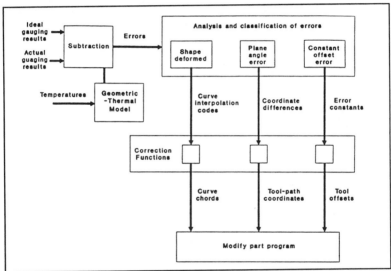

Figure 13

VL   POST-PROCESS INSPECTION USING DMIS AND IGES TO INTEGRATE CAD/CAM SOFTWARE PRODUCTS WITH CMMS

Post-process inspection monitors and provides the data to measure the
performance of, and improve, other parts of the QIA system, such as process-
intermittent (PI) part probing and the real-time error compensation algorithm
provided by the machine tool's geometric-thermal (G-T) error model. The
inspection procedure performed in this loop also determines whether a
finished part meets dimensional specifications.

Advances in production methods such as CNC machine tools and
integrated CAD/CAM products have increased the speed of the production

run. These increases in production speed have dictated the need for advances in part inspection techniques. This problem has been addressed by the use of coordinate measuring machines (CMMs). Not only have CMMs greatly reduced inspection times, but they have also increased the flexibility, reliability, and accuracy of measurements.

A. The Dimensional Measuring Interface Specification

In order to minimize the amount of time required for operator-CMM interfacing, a standard interface language is needed. In February, 1990, the American National Standards Institute (ANSI) adopted the Dimensional Measuring Interface Specification (DMIS) as a national standard [46]. DMIS is designed to prescribe a standard format in terms of a common language to allow interfacing between CAD/CAM software products and dimensional measuring equipment hardware products, such as CMMs.

DMIS serves as a bi-directional communication standard for computer systems and inspection equipment. In other words, inspection programs and the resulting inspection data can be expressed in a generic vocabulary that is available to both CAD/CAM software and CMMs and serves as the common link between the two. This allows the generation of part programs within CAM software that utilize data bases produced within a CAD package. These part programs can be generated off-line through graphical computer interfacing and the use of pre- and post-processors. This method, detailed in this text, greatly reduces the amount of time required to produce a part program.

B. System Overview - NIST Inspection System

An inspection system has been implemented at NIST that integrates CAD/CAM software products with a CMM using the DMIS standard. This system also utilizes the Initial Graphics Exchange Specification (IGES) to allow for interfacing between the different CAD and CAM software products that are present [47]. IGES provides for the exchange of graphical data among CAD systems from different manufacturers. The NIST system has been implemented with various goals in mind: (1) performing inspection for the post-process control loop in the QIA process; (2) demonstrating the operation of DMIS using products commercially available to the industrial public; (3) incorporating off-line automated inspection path generation directly from a CAD data base to the CMM using DMIS; (4) demonstrating the utility and flexibility of ANSI standards by integrating into the system products from many different manufacturers; and (4) addressing the needs of the small-to-moderate size machine shops in terms of cost and system simplicity by integrating the entire post-process inspection system into one enhanced PC-class computer with an 80386 processor.

The schematic flow diagram for the NIST automated inspection system is shown in Figure 14. This diagram illustrates graphically what the system does and how the system works. The realization of the strategy for this system is the direct result of a cooperative research effort among NIST and six U.S. industrial manufacturers, in conjunction with NIST's research associate (RA) program [48]. It is stressed that each module in Figure 14 represents a class of products that can be used in this type of system configuration; specific products are mentioned in the discussion of NIST's system with the intention of conveying general information regarding the integration and implementation of this type of system.

1. Inspection Process

The inspection process commences with the generation of a CAD representation of the manufactured part. This CAD representation is in the form of a three-dimensional wire-frame model and is produced within our system using a CAD package. The CAD data base is then translated into the IGES format using an IGES translator, to allow the CAD geometry to be utilized elsewhere in the system. The IGES standard provides a common graphics format that allows part geometry information to be used by various software products. This IGES geometry file is imported into a CAM software package, where the CMM inspection path is created in DMIS format. The path is created through a graphical interface on the computer screen, off-line and independent from the CMM. The user produces the inspection path by directing a mouse around the wire-frame image (IGES graphical representation) of the part geometry to designate the various points and features to be included in the part inspection. Once all the points, features, and other inspection information such as tolerance requirements and probe(s) data have been identified, the inspection routine, which is generated by the system, can be simulated on the computer. This is basically an editing feature of the system, and when all input and editing is completed, the software produces a DMIS part program. This DMIS part program is then translated into a part program that the CMM can execute using commercial translators. The inspections results are also stored in DMIS format for future analysis.

2. Inspection of Parts

There are two different strategies using DMIS for part inspections. For parts containing basic geometric features (circles, lines, etc.) the usual procedure would be to use commands to evaluate these features. In this case the points measured on the part have been processed to yield meaningful entities such as a circle diameter. Alternatively, an inspection program could be written containing commands which produce raw point data, in the form of x,y, and z coordinates, as the final result. Contoured parts generally do not contain easily identifiable geometric features; therefore, these parts must be inspected in this manner, with the aim of collecting information from a number of individual points that constitute a certain surface or curve on the part.

Parts can be inspected using part programs that contain measurement functions from the vocabulary of DMIS commands for basic geometric features, or the part programs can utilize the DMIS commands for general curves and surfaces, "FEAT/GCURVE" and "FEAT/GSURF," respectively. The specific set of DMIS commands used in the part program dictates the format of the inspection results, as well as the result analysis. For example,

TABLE 4. DMIS Vocabulary of Feature Definitions

FEAT/ARC   Format 1
FEAT/ARC   Format 2
FEAT/CIRCLE
FEAT/CONE
FEAT/CYLNDR
FEAT/ELLIPS
FEAT/GCURVE*
FEAT/GSURF*
FEAT/LINE
FEAT/PARPLN
FEAT/PATERN
FEAT/PLANE
FEAT/POINT
FEAT/RAWDAT
FEAT/RCTNGL
FEAT/SPHERE

   *   NOTE: These feature definitions are intended for use with contoured surfaces, rather than prismatic features.

the DMIS command to define a 2-D circle, "FEAT/CIRCLE," generates a measurement output that can include the coordinates of the circle center point and the diameter of the circle, among other entities. The DMIS command that defines a feature as a general curve, "FEAT/GCURVE," on the other hand, generates a measurement output that contains the 3-D coordinates of each point measured on the feature.   Contoured surface parts must be inspected on a point-by-point basis using the DMIS command for individual point identification, "FEAT/POINT," or using the DMIS "FEAT/GCURVE" or "FEAT/GSURF" commands. All three of the DMIS commands just mentioned produce a DMIS output file of inspection results that contains the x, y, and z coordinate values of each point measured on the part.

3. DMIS Part Programs

The inspection system at NIST can produce DMIS part programs for prismatic parts that implement the commands for features from the DMIS vocabulary (see Table 4). For contoured surface parts our system has the capability to generate DMIS programs that produce individual data point output only using the "FEAT/POINT" function.   Within our system we have the capability to conduct detailed analyses of such individual data point results, both graphically and numerically through algorithmic fitting routines, using the Qualstar metrology-based analysis software. At this time, splines representing contoured surfaces can be directly imported into Qualstar from their IGES entities, while simple geometric entities can be constructed in the software using an on-board editor.

4. Analysis of Results

For most prismatic parts either of the two inspection result formats described above provide adequate inspection information. However, for sculptured or contoured surfaces, DMIS results provide only x, y, and z probe coordinate data. Analysis of this type of data is performed within a commercially available software package.

The analysis of inspection results is carried out by comparing nominal part geometry with the actual, inspected part geometry. If the nominal part geometry is defined by the IGES parametric spline surface entities, the IGES file created earlier in the process may be used. Otherwise, the nominal geometry must be re-created. The analysis here is conducted on a point-by-point and feature-by-feature basis for a part, and mainly consists of graphically overlaying inspection data onto nominal part data. Numerous fitting routines and geometric algorithms are available in this software that allow data manipulation and analysis capabilities. In addition to indicating the out-of-tolerance parts, the analysis provides information about the possible reason for the part rejection by examining every data point in an effort to depict trends in the manufacturing process.

5. Limitations

The limitations of the inspection system can be divided into two categories: fundamental and application-dependent. A fundamental limitation of this system is set by the scope of the various national standards. IGES version 4.0 does not associate tolerance information with geometric features. Thus, it is necessary to input the tolerances during the inspection path generation or analysis steps. The Standard for the Exchange of Product Model Data (STEP, and formerly known as PDES), which will eventually supersede IGES, will directly associate the tolerances with their geometric features [49]. This would allow the tolerance information to come directly from the CAD file and increase the efficiency of the system.

The scope of the DMIS standard supported by the various application programs represents a more immediate limitation. Since the DMIS standard is intended to be applicable to a wide range of automated dimensional measuring equipment, few applications support all the statements within the standard. Consequently, a particular application program may not contain a desired command or definition. In our system, an example of this is the inability to calibrate or change the CMM probe directly from DMIS commands. While the system can make allowances for different positions of a probe by redefining the probe index within the part program, DMIS has no means of changing the physical type of probe sensor the CMM utilizes. In other words, DMIS has the capability to manipulate one probe head into many different probing angles within a part program, but no way to actually change, for example, that probe from a mechanical touch-trigger probe to a touch-trigger probe of different tip radius, or to a piezoelectric probe, should one of these be required. The NIST system does have a Renishaw automatic probe changer rack that allows the CMM physically to change the type of probe it uses. In practice, we execute a short calibration program written in the CMM's native language before starting the CMM inspection run. Since the probe calibration program is relatively short, and does not usually change between part inspections, in contrast to part programs which presumably change often, this is only a minor inconvenience. Similarly a probe can be exchanged in the auto change rack, but this command is in the native CMM language.

VII. CONCLUSION

Development of a closed-loop quality control system for discrete part manufacturing is one of the most challenging research efforts in the manufacturing field. With the Quality In Automation (QIA) program, a three-layered control architecture has been constructed to address this need and the steps required to demonstrate the inner two control layers have been completed. Even though the control architecture described here has been successfully implemented on a specific group of machine tools, the principles articulated herein are generally applicable to other discrete-part manufacturing systems. These systems would include a wide variety of machines, such as grinding machines and diamond turning machines.

In addition, the process is applicable to non-machining process such as the fabrication of weldments.

As U. S. industry reduces inventories by just-in-time delivery, traditional statistical process control cannot be used to control the quality of the parts produced. To further complicate matters, customers are demanding a higher quality part which translates to a higher accuracy part. The principles described herein can meet these demands by making the existing machining process more accurate. We have demonstrated an increase in accuracy of twenty by correcting the geometric and thermal errors on a turning center. We will continue to extend this effort to other machining processes so that we can achieve the same increase in accuracy every time for lot sizes of one.

This program is a combined effort of two divisions in the Manufacturing Engineering Laboratory at NIST. It is funded primarily by the U.S. Navy and internal NIST resources.

REFERENCES

1. G. Schlesinger, "Testing Machine Tools," 8th Edition, Pergamon Press, 1978.

2. J. Tlusty, "Techniques for Testing Accuracy of NC Machine Tools," "Proc. of 12th MTDR Conf.," 1971.

3. D. L. Leete, "Automatic Compensation of Alignment Errors in Machine Tools," "Int. Journal of MTDR," Vol. 1, 1961.

4. G. S. K. Wong and F. Koenigsberger, "Automatic Correction of Alignment Errors in Machine Tools," "Int. Journal of MTDR," Vol. 6, 1967.

5. T. C. Goodhead et al., "Automatic Detection of and Compensation for Alignment Errors in Machine Tool Slideways," "Proc. of 18th MTDR Conf.," 1977.

6. J. B. Bryan and D. L. Carter, "Design of a New Error-Corrected Coordinate Measuring Machine," "Precision Engineering," Vol. 1, 1979.

7. K. Okushima, Y. Kakino and A. Higashimoto, "Compensation of Thermal Displacement by Coordinate System Correction," "Annals of CIRP," Vol. 24, 1975.

8. T. Sata, Y. Takeuchi and N. Okubo, "Improvement in the Working Accuracy of Machining Center by Computer Control Compensation," "Proc. of 17th MTDR Conf.," 1976.

9. Y. Takeuchi, M. Sakomoto and T. Sata, "Improvement in the Working Accuracy of an NC Lathe by Compensating for Thermal Expansion," "Precision Engineering," Vol. 4, No. 1, 1982.

10. A. S. Koliskor et. al., "Compensating for Automatic-Cycle Machining Errors," "Machines and Tooling," Vol. 41, No. 5, 1971.

11. M. A. Donmez, C. R. Liu, and M. M. Barash, "A Generalized Mathematical Model for Machine-Tool Errors," "Modeling, Sensing, and Control of Manufacturing Processes6," PED-Vol. 23/DSC-Vol. 4, Book No. H00370, K. Srinivasan, D. L. E. Hardt, and R. Komanduri, eds. (The American Society of Mechanical Engineers, New York, 1987).

12. R. P. Paul, "Robot Manipulators: Mathematics, Programming, and Control," The MIT Press, 1981.

13. L. M. Chao and J. C. S. Yang, "Implementation of a Scheme to Improve the Positioning Accuracy of an Articulate Robot by using Laser Distance-Measuring Interferometry," "Proceeding of the 4th International Precision Engineering Seminar," 11-14 May 1987.

14. C. S. G. Lee, "Robot Arm Kinematics," "Tutorial on Robotics," IEEE Computer Society, 1983.

15. R. P. Paul, B. Shimano and G. E. Mayer, "Kinematic Control Equations for Simple Manipulators," "Tutorial on Robotics," IEEE Computer Society, 1983.

16. J. Denavit and R. S. Hartenberg, "A Kinematic Notation for Lower-Pair Mechanisms Based on "Journal of Applied Mechanics," June 1955.

17. J. Tlusty and F. Koeningsberger, "New Concepts of Machine Tool Accuracy," "Annals of CIRP", 1971.

18. C. P. Hemingway, "Some Aspects of the Accuracy Evaluation of Machine Tools," "Proc. of 14th MTDR Conf.," 1973.

19. H. Sato, "Machine Tool," "Bull. of Japan Soc. of Precision Eng.," Vol. 8, No. 2, 1974.

20. B. M. Bazrov, "Investigating Machining Accuracy Using a Computer," "Machines and Tooling," Vol. 47, No. 8, 1976.

21. R. J. Hocken, "Quasistatic Machine Tool Errors," "Technology of Machine Tools," MTTF, Vol. 5, 1980.

22. Y. Takeuchi and M. Sakamoto, "Analysis of Machining Error in Face Milling," "Proc. of 22nd MTDR Conf.," 1981.

23. M. A. Donmez et. al., "Statistical Analysis of Positioning Error of a CNC Milling Machine," "Journal of Manufacturing Systems," Vol. 1, No. 1, 1982.

24. K. L. Blaedel, "Error Reduction, "Technology of Machine Tools," MTTF, Vol. 5, 1980.

25. M. A. Donmez, D. S. Blomquist, R. J. Hocken, C. R. Liu, and M. M. Barash, "A General Methodology for Machine Tool Accuracy Enhancement by Error Compensation," "Precision Engineering," Vol. 4, 1986.

26. F. W. Jones, "Performance Evaluation of Precision Numerically Controlled Turning Equipment," "Proc. of MTDR Conf.," 1970.

27. V. L. Romanov and O. P. Philimonovskaya, "Checking the Accuracy of the Coordinate Displacement of NC Machine Tools," "Machines and Tooling," Vol. 48, No. 3, 1977.

28. V. T. Portman et. al., "Investigating the Positioning Accuracy of a NC Contour Grinder," "Machines and Tooling," Vol. 50, No. 6, 1979.

29. R. Schultschick, "The Components of Volumetric Accuracy," "Annals of CIRP," Vol. 26, 1977.

30. V. T. Portman, "Error Summation in the Analytical Calculation of Lathe Accuracy," "Machines and Tooling," Vol. 51, No. 1, 1980.

31. M. I. Koval and G. A. Igonin, "Comparative Analysis of Machining Error Components for a Heavy NC Machine Tool," "Machines and Tooling," Vol. 50, No. 9, 1979.

32. R. J. Hocken et. al., "Three Dimensional Metrology," "Annals of CIRP," Vol. 26, 1977.

33. E. R. McClure, "Manufacturing Accuracy Through the Control of Thermal Effects," Ph. D. Thesis, University of California Livermore.

34. R. L. Murty, "Thermal Deformation of a Semi-automatic Machine: A Case Study," "Precision Engineering," Vol. 2, No. 1, 1980.

35. M. H. Attia and L. Kops, "System Approach to the Thermal Behavior and Deformation of Machine Tool Structures in Response to the Effect of Fixed Joints," "Transactions of ASME," 80-WA/PROD-14.

36. G. Spur and P. DeHaas, "Thermal Behavior of NC Machine Tools," "Proc. of 14th MTDR Conf.," 1973.

37. E. R. McClure, "Thermally Induced Errors," "Technology of Machine Tools," MTTF, Vol. 5, 1980.

38. M. H. Attia and L. Kops, "Concept of Thermoelastic Interactions at Machine Tool Joints - A Basis for Improved Machining Accuracy," "Manufacturing Solutions Based on Engineering Sciences," ASME, 1981.

39. J. Tlusty and G. F. Mutch, "Testing and Evaluating Thermal Deformations of Machine Tools," "Proc. of 14th MTDR Conf.," 1973.

40. T. Sata, Y. Takeuchi and N. Okubo, "Analysis of Thermal Deformation of Machine Tool Structure and Its Application," "Proc. of 14th MTDR Conf.," 1973.

41. H. T. Bandy, V. E. Carew, Jr., and J. C. Boudreaux, "An AMPLE Version 0.1 Prototype: The HWS Implementation," National Bureau of Standards (US), NBSIR 88-3770, April 1988.

42. J. C. Boudreaux, "AMPLE: A Programming Language Environment for Automated Manufacturing," "The Role of Language in Problem Solving - 2," J. C. Boudreaux, B. Hamill, and R. Jernigan eds., (North Holland, Amsterdam, 1987).

43. "Progress Report of the Quality in Automation Project for FY88," C. Denver Lovett, ed., National Institute of Standards and Technology (US), NISTIR 89-4045, April 1989.

44. "Progress Report of the Quality In Automation Project for FY89," T. V. Vorburger and B. Scace, eds., National Institute of Standards and Technology (US), NISTIR 4322, May 1990.

45. "Progress Report of the Quality in Automation Project for FY90," M. A. Donmez, ed., National Institute of Standards and Technology (US), NISTIR 4536, March 1991.

46. "DMIS 2.1," Specification CAM-I Standard 101 (CAMI, Inc., Arlington, TX, 1990).

47. Initial Graphics Exchange Specification (IGES) Version 4.0, National Bureau of Standards (US), NBSIR 88-3813, June 1988.

48. NIST Research Associate Program information folder (U.S. Department of Commerce, National Institute of Standards and Technology, Gaithersburg, MD, 1988).

49. B. Smith, "Product Data Exchange: The PDES Project - Status and Objectives," National Institute of Standards and Technology (US), NISTIR 89-4165, September 1989.

A THEORY OF
INTELLIGENT SYSTEMS

JAMES S. ALBUS

Robot Systems Division
Manufacturing Engineering Laboratory
National Institute of Standards and Technology
Gaithersburg, MD   20899

I. INTRODUCTION

Much is unknown about intelligence, and much will elude human understanding for a very long time.   Yet much is known, both about the mechanisms and function of intelligence. It is not too soon to propose at least the beginnings of a theory of intelligent systems.

The study of intelligent machines and the neurosciences are extremely active fields. Many millions of dollars per year are now being spent in Europe, Japan, and the United States on computer integrated manufacturing, robotics, and intelligent machines for a wide variety of military and commercial applications. Around the world, researchers in the neurosciences are searching for the anatomical, physiological, and chemical basis of behavior. Research in learning automata, neural nets, and brain modeling has given insight into learning and the similarities and differences between neuronal and electronic computing processes.   Computer science and artificial intelligence are probing the nature of language and image understanding, and have made significant progress in rule based reasoning, planning, and problem solving.   Game theory and operations research have developed methods for decision making in the face of uncertainty.   Robotics and autonomous vehicle research have produced advances in real-time sensory processing, world modeling, navigation, trajectory generation,   and obstacle avoidance. Research in automated manufacturing and process control has produced intelligent hierarchical controls, distributed databases, representations of object geometry and material properties, data driven task sequencing, network communications, and multiprocessor operating systems. Modern control theory has developed precise understanding of stability, adaptability, and controllability under various conditions of feedback and noise. Research in sonar, radar, and optical signal processing has developed methods for fusing sensory input from multiple sources, and assessing the believability of noisy data.

Progress is rapid, and there exists an enormous and rapidly growing literature in each of the areas mentioned above. What is lacking is a general theoretical model of intelligent systems which ties all these separate fields of knowledge into a unified framework. This paper is an attempt to begin the construction of such a framework.

II. THE ELEMENTS OF INTELLIGENCE

The elements of intelligence and their relationship to each other are illustrated in Figure 1 and described below.

A. Actuators

Output from an intelligent system derives from actuators which move, exert forces, and position arms, legs, hands, and eyes. Actuators generate forces to point sensors, excite transducers, move manipulators, handle tools, steer and propel locomotion. An intelligent system may have tens, hundreds, or even thousands of actuators, all of which must be coordinated in order to perform tasks and accomplish goals. Natural actuators are muscles and glands. Machine actuators are motors, pistons, valves, solenoids, and transducers.

B. Sensors

Input to an intelligent system derives from sensors. These may include visual brightness and color sensors; tactile, force, torque, position detectors; velocity, vibration, acoustic, range, smell, taste, pressure, and temperature measuring devices. Sensors may be used to monitor both the state of the external world and the internal state of the intelligent system itself. Sensors provide input to a sensory processing system.

C. Sensory Processing

Perception takes place in a sensory processing system that compares observations with expectations generated by an internal world model. Sensory processing algorithms integrate similarities and differences between observations and expectations, over time and space, so as to detect events, and recognize features, objects, and relationships in the world. Input data from a wide variety of sensors over extended periods of time may be fused into a consistent unified perception of the state of the world. Sensory processing algorithms compute distance, shape, orientation, surface characteristics, physical and dynamical attributes of objects and regions of space. Sensory processing may include recognition of acoustic signatures, speech, and interpretation of language.

D. World Model

The world model is the intelligent system's best estimate of the state of the world. The world model includes a database of knowledge about the world, plus a database management system that stores and retrieves information. The world model also contains a simulation capability which generates expectations and predictions. The world model thus can provide answers to requests for information about the present, past, and probable future states of the world. The world model provides this information service to the task decomposition system, so that it can make intelligent plans and behavioral choices, and to the sensory processing system, in order for it to perform correlation, model matching, and model based recognition of states, objects, and events. The world model is kept up-to-date by the sensory processing system.

E. Values

The value system makes value judgments as to what is good and bad, rewarding and punishing, important and trivial. The value system evaluates both the observed state of the world and the predicted results of hypothesized plans. It computes costs, risks, and benefits both of observed situations and of planned activities. The value system thus provides the basis for choosing one action as opposed to another, or for acting on one object as opposed to another.   The value system also computes the probability of correctness and assigns believability and uncertainty parameters to world model state estimations.

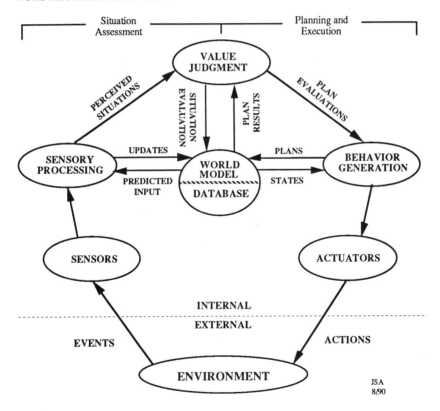

Figure 1.   The elements of intelligence and the functional relationships between them.

F. Task Decomposition

Behavior is generated in a task decomposition system that plans and executes tasks by decomposing them into subtasks, and by sequencing these subtasks so as to achieve goals. Goals are selected and plans generated by a looping interaction between task decomposition, world modeling, and value judgment functions. The task decomposition system hypothesizes plans, the world model predicts the results of those plans, and the value judgment system evaluates those results. The task decomposition system then selects the plans with the best evaluations for execution. Task decomposition monitors the execution of task plans, and modifies existing plans whenever the situation requires.

In many cases, intelligent task decomposition requires the ability to reason about space and time, geometry and dynamics, and to formulate or select plans based on values such as cost, risk, utility, and goal priorities. Task planning and execution often must be done in the presence of uncertain, incomplete, and sometimes incorrect information.

In order for task decomposition to succeed in a dynamic and unpredictable world, it must be accomplished in real-time. In order to achieve real-time task decomposition, it is necessary to partition the planning problem into a hierarchy of levels with different temporal planning horizons and different degrees of detail at each hierarchical level. Once this is done, it is possible to employ a multiplicity of planners to simultaneously generate and coordinate plans for many different subsystems at many different levels.

III. THE SYSTEM ARCHITECTURE OF INTELLIGENCE

Each of the elements of intelligent systems are reasonably well understood. The phenomena of intelligence, however, require more than a set of disconnected elements. Intelligence requires an interconnecting system architecture that enables the various system components to interact and communicate with each other in intimate and sophisticated ways.

A system architecture is what partitions the elements of intelligence into computational modules, and interconnects the modules in networks and hierarchies. It is what enables the task decomposition system to direct sensors, and to focus sensory processing algorithms on objects and events worthy of attention, ignoring things that are not important to current goals and task priorities. It is what enables the world model to answer queries from task decomposition modules, and make predictions and receive updates from sensory processing modules. It is what communicates the value state-variables that characterize the success of behavior and the desirability of states of the world.

A number of intelligent system architectures have been proposed, and a few have been implemented [1-10]. The model of intelligence that will be discussed here is largely based on the Real-time Control System (RCS) that has been implemented in a number of versions over the past 13 years at the National Institute of Standards and Technology (NIST formerly NBS). RCS was first implemented by Barbera for laboratory robotics in the mid 1970's [7] and adapted by Albus, Barbera, and others for manufacturing control in the NIST Automated Manufacturing Research Facility (AMRF) during the early 1980's [11,12]. Since 1986, RCS has been implemented for a number of additional applications, including the NBS/DARPA Multiple Autonomous Undersea Vehicle (MAUV) project [13] and the Army TMAP and TEAM semi-autonomous land vehicle projects. RCS also forms the basis of the NASA/NBS Standard Reference Model Telerobot Control System Architecture (NASREM) being used on the space station Flight Telerobotic Servicer [14].

The proposed system architecture organizes the elements of intelligence so as to create the functional relationships and information flow shown in Figure 1. In all intelligent systems, a sensory processing system processes sensory information to acquire and maintain an internal model of the external world. In all systems, a behavior generating system controls actuators so as to pursue behavioral goals in the context of the perceived world model. In systems of higher intelligence, the behavior generating system element may interact with the world model and value judgment system to reason about space and time, geometry and dynamics, and to formulate or select plans based on values such as cost, risk, utility, and goal priorities. The sensory processing system element may interact with the world model and value judgment system to assign values to perceived entities, events, and situations.

The proposed system architecture replicates and distributes the relationships shown in Figure 1 over a hierarchical computing structure with the logical and temporal properties illustrated in Figure 2. On the left is an organizational hierarchy wherein computational nodes are arranged in layers like command posts in a military organization. Each node in the organizational hierarchy contains four types of computing modules: behavior generating (BG), world modeling (WM), sensory processing (SP), and value judgment (VJ) modules. Each chain of command in the organizational hierarchy, from each actuator and each sensor to the highest level of control, can be represented by a computational hierarchy, such as is shown in the center of Figure 2.

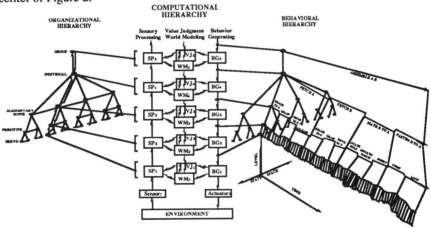

Figure 2.   Relationships in hierarchical control systems. On the left is an organizational hierarchy consisting of a tree of command centers, each of which possesses one supervisor and one or more subordinates. In the center, is a computational hierarchy consisting of BG, WM, SP, and VJ modules. Each actuator and each sensor is serviced by a computational hierarchy. On the right is a behavioral hierarchy consisting of trajectories through state-time-space.   Commands at each level can be represented by vectors, or points in state-space. Sequences of commands can be represented as trajectories through state-time-space.

At each level, the nodes, and computing modules within the nodes, are richly interconnected to each other by a communications system. Within each computational node, the communication system provides intermodule communications of the type shown in Figure 1. Queries and task status are communicated from BG modules to WM modules. Retrievals of information are communicated from WM modules back to the BG modules making the queries. Predicted sensory data is communicated from WM modules to SP modules. Updates to the world model are communicated from SP to WM modules. Observed entities, events, and situations are communicated from SP to VJ modules. Values assigned to the world model representations of these entities, events, and situations are communicated from VJ to WM modules. Hypothesized plans are communicated from BG to WM modules. Results are communicated from WM to VJ modules. Evaluations are communicated from VJ modules back to the BG modules that hypothesized the plans.

The communications system also communicates between nodes at different levels. Commands are communicated downward from supervisor BG modules in one level to subordinate BG modules in the level below. Status reports are communicated back upward through the world model from lower level subordinate BG modules to the upper level supervisor BG modules from which commands were received. Observed entities, events, and situations detected by SP modules at one level are communicated upward to SP modules at a higher level. Predicted attributes of entities, events, and situations stored in the WM modules at a higher level are communicated downward to lower level WM modules. Output from the bottom level BG modules is communicated to actuator drive mechanisms. Input to the bottom level SP modules is communicated from sensors.

The communications system can be implemented in a variety of ways. In a biological brain, communication is mostly via neuronal axon pathways, although some messages are communicated by hormones carried in the bloodstream. In artificial systems, the physical implementation of communications functions may be a computer bus, a local area network, a common memory, a message passing system, or some combination thereof. In either biological or artificial systems, the communications system may include the functionality of a communications processor, a file server, a database management system, a question answering system, or an indirect addressing or list processing engine. In the system architecture proposed here, the input/output relationships of the communications system produce the effect of a virtual global memory, or blackboard system [15].

The string of input commands to each of the BG modules at each level generates a trajectory through state-space as a function of time. The set of command strings to all BG modules creates a behavioral hierarchy, as shown on the right of Figure 2. Actuator output trajectories (not shown in Figure 2) correspond to observable output behavior. All the other trajectories in the behavioral hierarchy constitute the deep structure of behavior [16].

A. Hierarchical VS. Horizontal

Figure 3 shows the organizational hierarchy in more detail, and illustrates both the hierarchical and horizontal relationships involved in the proposed architecture. The architecture is hierarchical in that commands and status feedback flow hierarchically up and down a behavior generating chain of command. The architecture is also hierarchical in that sensory processing and world modeling functions have hierarchical levels of temporal and spatial aggregation. The command hierarchy is a tree, in that at any instant of time any BG module has only one supervisor.

The SP modules are also organized hierarchically, but as a layered graph, not a tree. At each higher level, sensory information is processed into increasingly higher levels of abstraction, but the sensory processing pathways may branch and merge in many different ways.

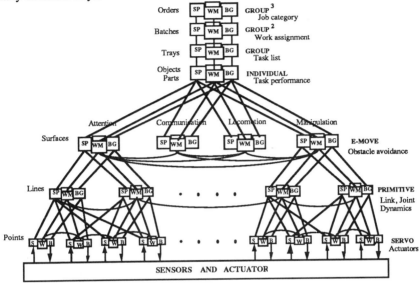

Figure 3. An organization of processing nodes such that the BG modules form a command tree. On the right are examples of the functional characteristics of the BG modules at each level. On the left are examples of the type of visual and acoustical entities operated on by the SP modules at each level. In the center of level 3 are the type of subsystems represented by processing nodes at level 3.

The architecture is horizontal in that data is shared horizontally between heterogeneous modules at the same level. At each hierarchical level, the architecture is horizontally interconnected by wide-bandwidth communication pathways between BG, WM, SP, and VJ modules in the same node, and between nodes at the same level, especially within the same command subtree. The horizontal flow of information is most voluminous within a single node, less so between related nodes in the same command subtree, and relatively low bandwidth between computing modules in separate command subtrees. Communications bandwidth is indicated in Figure 3 by the thickness of the horizontal connections.

The volume of information flowing horizontally within a subtree may be orders of magnitude larger than the amount flowing vertically in the command chain. The volume of information flowing vertically in the sensory processing system can also be very high, especially in a vision system.

The specific configuration of the command tree is task dependent, and therefore not necessarily stationary in time. Figure 3 illustrates only one possible configuration that may exist at a single point in time. During operation, relationships between modules within and between layers of the hierarchy may be reconfigured in order to accomplish different goals, priorities, and task requirements. This means that any particular computational node, with its BG, WM, SP, and VJ modules, may belong to one subsystem at one time and a different subsystem a very short time later. For example, a mobile robot may belong to one workstation at one instant in time, and a different workstation a very short time later. Similarly, a robot may employ one gripper for one task, and a different gripper with different capabilities for another task.

In the biological brain, command tree reconfiguration can be implemented through multiple axon pathways that exist, but are not always activated, between BG modules at different hierarchical levels. In intelligent machine systems, multiple communications pathways may enable each BG module to receive input messages and parameters from several different sources. During operation, goal driven switching mechanisms in the BG modules assess priorities, negotiate for resources, and coordinate task activities so as to select among the possible communication paths. As a result, each BG module will accept task commands from only one supervisor at a time, and hence the BG modules form a command tree at every instant in time.

B. Hierarchical Levels

Levels in the behavior generating hierarchy are defined by temporal and spatial decomposition of goals and tasks into levels of resolution. Temporal resolution is manifested in terms of loop bandwidth, sampling rate, and state-change intervals. Temporal span is measured by the length of historical traces and planning horizons. Spatial resolution is manifested in the branching of the command tree and the resolution of maps. Spatial span is measured by the span of control and the range of maps.

Levels in the sensory processing hierarchy are defined by temporal and spatial integration of sensory data into levels of aggregation. Spatial aggregation is best illustrated by visual images. Temporal aggregation is best illustrated by acoustic parameters such as phase, pitch, phonemes, words, sentences, rhythm, beat, and melody.

Levels in the world model hierarchy are defined by temporal resolution of events, spatial resolution of maps, and by "parent-child" relationships between entities in symbolic data structures. These are defined by the needs of both SP and BG modules at the various levels.

Hypothesis: In a hierarchically structured goal-driven, sensory-interactive, intelligent control system architecture:

a)   control bandwidth decreases about an order of magnitude at each higher level,
b)   perceptual resolution of spatial and temporal patterns decreases about an order-of-magnitude at each higher level,
c)   goals expand in scope and planning horizons expand in space and time about an order-of-magnitude at each higher level, and
d)   models of the world and memories of events decrease in resolution and expand in spatial and temporal range by about an order-of-magnitude at each higher level.

It is well known from control theory that hierarchically nested servo loops tend to suffer instability unless the bandwidth of the control loops differ by about an order of magnitude. This suggests, perhaps even requires, condition a) above. Numerous theoretical and experimental studies support the concept of hierarchical planning and perceptual "chunking" for both temporal and spatial entities [17,18].

These support conditions b), c), and d) above.

In elaboration of the above hypothesis, we can construct a timing diagram, as shown in Figure 4. The range of the time scale increases, and its resolution decreases, exponentially by about an order of magnitude at each higher level. Hence the planning horizon and event summary interval increases, and the loop bandwidth and frequency of subgoal events decreases, exponentially at each higher level. The seven hierarchical levels in Figure 4 span a range of time intervals from three milliseconds to one day. Three milliseconds was arbitrarily chosen as the shortest servo update rate because that is adequate to reproduce the highest bandwidth reflex arc in the human body. One day was arbitrarily chosen as the longest historical-memory/planning-horizon to be considered. Shorter time intervals could be handled by adding another layer at the bottom. Longer time intervals could be treated by adding layers at the top, or by increasing the difference in loop bandwidths and sensory chunking intervals between levels.

The origin of the time axis in Figure 4 is the present, i.e. t=0. Future plans lie to the right of t=0, past history to the left. The open triangles in the right half-plane represent task goals in a future plan. The filled triangles in the left half-plane represent recognized task-completion events in a past history. At each level there is a planning horizon and a historical event summary interval. The heavy cross-hatching on the right shows the planning horizon for the current task. The light shading on the right indicates the planning horizon for the anticipated next task. The heavy cross-hatching on the left shows the event summary interval for the current task. The light shading on the left shows the event summary interval for the immediately previous task.

Figure 4 suggests a duality between the behavior generation and the sensory processing hierarchies. At each hierarchical level, planner modules decompose task commands into strings of planned subtasks for execution. At each level, strings of sensed events are summarized, integrated, and "chunked" into single events at the next higher level.

Planning implies an ability to predict future states of the world. Prediction algorithms based on Baysian statistics, Fourier transforms, or Kalman filters typically use recent historical data to compute parameters for extrapolating into the future. Predictions made by such methods are typically not reliable for periods longer than the historical interval over which the parameters were computed. Thus at each level, planning horizons extend into the future only about as far, and with about the same level of detail, as historical traces reach into the past.

Predicting the future state of the world often depends on assumptions as to what actions are going to be taken and what reactions are to be expected from the environment, including what actions may be taken by other intelligent agents. Planning of this type requires search over the space of possible future actions and probable reactions. Search-based planning takes place via a looping interaction between the BG, WM, and VJ modules whereby various possible futures are simulated and evaluated. This is described in more detail in the section on BG modules.

Planning complexity grows exponentially with the number of steps in the plan (i.e. the number of layers in the search graph). If real-time planning is to succeed, any given planner must operate in a limited search space. If there is too much resolution in the time line, or in the space of possible actions, the size of the search graph can easily become too large for real-time response. One method of resolving this problem is to use a multiplicity of planners in hierarchical layers [14,19] so that at each layer no planner needs to search more than a given number (for example ten) steps deep in a game graph, and at each level there are no more than (ten) subsystem plans that need to be simultaneously generated and coordinated.

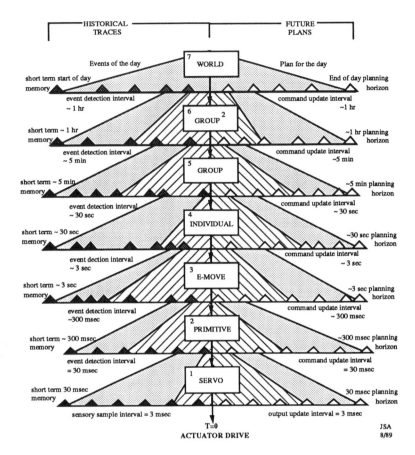

Figure 4. A timing diagram illustrating the temporal flow of activity in the task decomposition and sensory processing systems. At the world level, high level sensory events and circadian rhythms react with habits and daily routines to generate a plan for the day. Each element of that plan is decomposed through the remaining six levels of task decomposition into action.

These criteria give rise to hierarchical levels with exponentially expanding spatial and temporal planning horizons, and characteristic degrees of detail for each level. The result of hierarchical spatio-temporal planning is illustrated in Figure 5. At each level, plans consist of at least one, and on average 10, subtasks. The planners have a planning horizon that extends about one and a half average input command intervals into the future.

In a real-time system, plans must be regenerated periodically to cope with changing and unforeseen conditions in the world. Cyclic replanning may occur at periodic intervals. Emergency replanning begins immediately upon the detection of an emergency condition. Under full alert status, the cyclic replanning interval should be about an order of magnitude less than the planning horizon (or about equal to the expected output subtask time duration). This requires that real-time planners be able to search to the planning horizon about an order of magnitude faster than real time. This is possible only if the depth and resolution of search is limited through hierarchical planning.

Plan executors at each level have responsibility for reacting to feedback every control cycle interval. Control cycle intervals are inversely proportional to the control loop bandwidth. Typically the control cycle interval is an order of magnitude less than the expected output subtask duration. If the feedback indicates the failure of a planned subtask, the executor branches immediately (i.e. in one control cycle interval) to a preplanned emergency subtask. The planner simultaneously selects or generates an error recovery sequence which is substituted for the former plan which failed. Plan executors are also described in more detail in section 3.

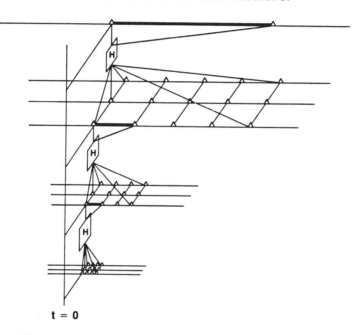

t = 0

Figure 5.   Three levels of real-time planning illustrating the shrinking planning horizon and greater detail at successively lower levels of the hierarchy. At the top level a single task is decomposed into a set of four planned subtasks for each of three subsystems. At each of the next two levels, the first task in the plan of the first subsystem is further decomposed into four subtasks for three subsystems at the next lower levels.

When a task goal is achieved at time t=0, it becomes a task completion event in the historical trace. To the extent that a historical trace is an exact duplicate of a former plan, there were no surprises; i.e. the plan was followed, and every task was accomplished as planned. To the extent that a historical trace is different from the former plan, there were surprises. The average size and frequency of surprises (i.e.

differences between plans and results) is a measure of effectiveness of a planner (indeed of an intelligent control system).

At each level in the control hierarchy, the difference vector between planned (i.e. predicted) and observed events is an error signal, that can be used by executor submodules for servo feedback control (i.e. error correction), and by VJ modules for evaluating success and failure.

In the next four sections, the system architecture outlined above will be elaborated and the functionality of the computational submodules for behavior generation, world modeling, sensory processing, and value judgment will be discussed.

C. Behavior

Behavior is the result of executing a series of tasks. A task is a piece of work to be done, or an activity to be performed For any intelligent system, there exists a set of tasks that the system knows how to do. Each task in this set can be assigned a name. The task vocabulary is the set of task names assigned to the set of tasks the system is capable of performing. For creatures capable of learning, the task vocabulary is not fixed in size. It can be expanded through learning, training, or programming. It may shrink from forgetting, or program deletion.

Typically, a task is performed by one or more actors on one or more objects. The performance of a task can usually be described as an activity which begins with a start-event and is directed toward a goal-event.

A goal may be an event which successfully terminates a task, or an objective toward which task activity is directed. A task command is an instruction to perform a named task. A task command may be expressed in a command frame [20] of the form:

TASKNAME   -- task identifier
goal   -- event which successfully terminates or
                renders the task successful
object   -- thing upon which task is to be performed
parameters -- priority
            -- status (e.g. active, waiting, inactive)
            -- timing requirements
            -- source of task command

Task knowledge is knowledge of how to perform a task, including information as to what tools, materials, time, resources, information, and conditions are required, plus information as to what costs, benefits and risks are expected.

Task knowledge may be expressed implicitly in fixed circuitry, either in the neuronal connections and synaptic weights of the brain, or in algorithms, software, and computing hardware. Task knowledge may also be expressed explicitly in data structures, either in the neuronal substrate or in a computer memory.

A task frame is a data structure in which task knowledge can be stored. In systems where task knowledge is explicit, a task frame [20] can be defined for each task in the task vocabulary. An example of a task frame is:

TASKNAME   -- task identifier
type   -- generic or specific
goal   -- event that successfully terminates or
                renders the task successful
object   -- thing upon which task is to be performed

parameters   --   priority
             --   status (e.g. active, waiting, inactive)
             --   timing requirements
             --   source of task command
actor   --   agent performing the task
requirements --   tools, time, resources, and materials
                needed to perform the task
             --   enabling conditions that must be satisfied
                to begin, or continue, the task
             --   disabling conditions that will prevent, or
                interrupt, the task
             --   information that may be required
procedures   --   a state-graph, state-table, or program
                defining a plan for executing the task
             --   functions that may be called
             --   algorithms that may be needed
effects   --   expected results of task execution
             --   expected costs, risks, benefits
             --   estimated time to complete

Note that the task name, goal, object, and parameters are duplicated in the task and command frames. This is because the command frame supplies these attributes to the task frame. Upon receipt of a command at execution time, the task vocabularly is searched for a task frame that has the same name as the command frame. That task frame is then instantiated as specific with the goal, object, and parameters carried by the command frame.

Explicit representation of task knowledge in task frames has a variety of uses. For example, task planners may use the procedure section for generating hypothesized actions. The world model may use the effects slots for predicting the results of hypothesized actions. The value judgment system may use the priority and requirements attributes for computing how important the goal is and how many resources are needed to achieve it. Plan executors may use the procedure plan for selecting what to do next.

Task knowledge is typically difficult to discover, but once known, can be readily transferred to others. Task knowledge may be acquired by trial and error learning, but more often it is acquired from a teacher, or from written or programmed instructions. For example, the common household task of preparing a food dish is typically performed by following a recipe. A recipe is an informal task frame for cooking. Gourmet dishes rarely result from reasoning about possible combinations of ingredients, still less from random trial and error combinations of food stuffs. Exceptionally good recipes often are closely guarded secrets that, once published, can easily be understood and followed by others.

Making steel is a more complex task example. Steel making took the human race many millennia to discover how to do. However, once known, the recipe for making steel can be implemented by persons of ordinary skill and intelligence.

In most cases, the ability to accomplish complex tasks successfully is more dependent on the amount of task knowledge stored in task frames (particularly in the procedure section) than on the sophistication of planners in reasoning about tasks.

D. Behavior Generation

Behavior generation is inherently a hierarchical process. At each level of the behavior generation hierarchy, tasks are decomposed into subtasks that become task commands to the next lower level. At each level of a behavior generation hierarchy there exists a task vocabulary and a corresponding set of task frames. Each task frame contains a procedure state-graph. Each node in the procedure state-graph must correspond to a task name in the task vocabulary at the next lower level.

Behavior generation consists of both spatial and temporal decomposition. Spatial decomposition partitions a task into jobs to be performed by different subsystems. Spatial task decomposition results in a tree structure, where each node corresponds to a BG module, and each arc of the tree corresponds to a communication link in the chain of command as illustrated in Figure 3.

Temporal decomposition partitions each job into sequential subtasks along the time line. The result is a set of subtasks, all of which when accomplished, achieve the task goal, as illustrated in   Figure 6.

In a plan involving concurrent job activity by different subsystems, there may requirements for coordination, or mutual constraints. For example, a start-event for a subtask activity in one subsystem may depend on the goal-event for a subtask activity in another subsystem. Some tasks may require concurrent coordinated cooperative action by several subsystems. Both planning and execution of subsystem plans may thus need to be coordinated.

There may be several alternative ways to accomplish a task. Alternative task or job decompositions can be represented by an AND/OR graph in the procedure section of the task frame. The decision as to which of several alternatives to choose is made through a series of interactions between the BG, WM, SP, and VJ modules. Each alternative may be analyzed by the BG module hypothesizing it, WM predicting the result, and VJ evaluating the result. The BG module then chooses the "best" alternative as the plan to be executed.

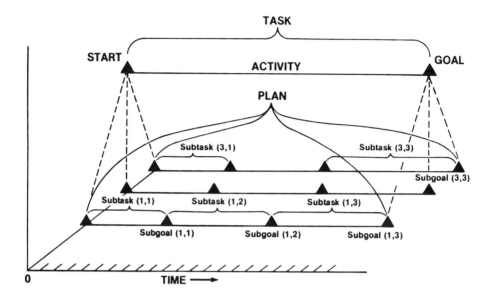

Figure 6. A task consists of an activity which typically begins with a start event and is terminated by a goal event. A task may be decomposed into several concurrent strings of subtasks which collectively achieve the goal event.

E. BG Modules

In the control architecture defined in Figure 3, each level of the hierarchy contains one or more BG modules. At each level, there is a BG module for each subsystem being controlled. The function of the BG modules are to decompose task commands into subtask commands.

Input to BG modules consists of commands and priorities from BG modules at the next higher level, plus evaluations from nearby VJ modules, plus information about past, present, and predicted future states of the world from nearby WM modules. Output from BG modules may consist of subtask commands to BG modules at the next lower level, plus status reports, plus "What Is?" and "What If?" queries to the WM about the current and future states of the world.

Each BG module at each level consists of three sublevels [9,14] as shown in Figure 7.

Task Decomposition

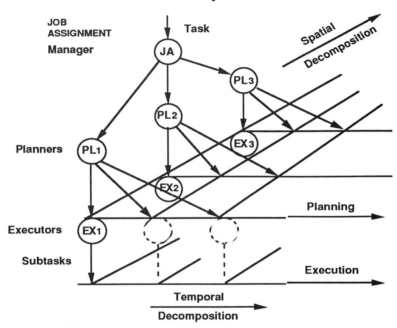

Figure 7. The job assignment JA module performs a spatial decomposition of the task. The planners PL(i) perform a temporal decomposition. The executors(i) execute the plans generated by the planners.

1) the job assignment sublevel -- JA submodule

The JA submodule is responsible for spatial task decomposition. It partitions the input task command into N spatially distinct jobs to be performed by N physically distinct subsystems, where N is the number of subsystems currently assigned to the BG module. The JA submodule may assign tools and allocate physical resources (such as arms, hands, mobility systems, sensors, tools, and materials) to each of its subordinate subsystems for their use in performing their assigned jobs. These assignments are not necessarily static. For example, the job assignment submodule at the cell level may assign a mobile robot to a machining workstation to load a tool or part into a machine tool, and later assign the same robot to a welding workstation to fixture a weldment or maneuver a welding torch.

The job assignment submodule selects the coordinate system in which the task decomposition at that level is to be performed. In supervisory or telerobotic control systems such as defined by NASREM [14], the JA submodule at each level may also determine the amount and kind of input to accept from a human operator.

2) the planner sublevel -- PL(j) submodules j = 1, 2, . . , N
   For each of the N subsystems, there exists a planner submodule PL(j). Each
planner submodule is responsible for decomposing the job assigned to its
subsystem into a temporal sequence of planned subtasks. Planner submodules
PL(j) may be implemented by case-based planners that simply select partially or
completely prefabricated plans, scripts, or schema [19,21,22] from the
procedure sections of task frames. This may be done by evoking situation/action
rules of the form, IF(case_x)/THEN(use_plan_y). The planner submodules
may complete partial plans by providing situation dependent parameters.
   The range of behavior that can be generated by a library of prefabricated plans
at each hierarchical level, with each plan containing a number of conditional
branches and error recovery routines, can be extremely large and complex. For
example, nature has provided biological creatures with an extensive library of
genetically prefabricated plans, called instinct. For most species, case-based
planning using libraries of instinctive plans has proven adequate for survival and
gene propagation in a hostile natural environment.
   Planner submodules may also be implemented by search-based planners that
search the space of possible actions. This requires the evaluation of alternative
hypothetical sequences of subtasks, as illustrated in Figure 8. Each planner
PL(j) hypothesizes some action or series of actions, the WM module predicts
the effects of those action(s), and the VJ module computes the value of the
resulting expected states of the world, as depicted in Figure 8(a). This results in
a game (or search) graph, as shown in 8(b). The path through the game graph
leading to the state with the best value becomes the plan to be executed by EX(j).
In either case-based or search-based planning, the resulting plan may be
represented by a state-graph, as shown in Figure 8(c). Plans may also be
represented by gradients, or other types of fields, on maps [23], or in
configuration space.
   Job commands to each planner submodule may contain constraints on time, or
specify job-start and job-goal events. A job assigned to one subsystem may also
require synchronization or coordination with other jobs assigned to different
subsystems. These constraints and coordination requirements may be specified
by, or derived from, the task frame. Each planner PL(j) submodule is
responsible for coordinating its plan with plans generated by each of the other
N-1 planners at the same level, and checking to determine if there are mutually
conflicting constraints. If conflicts are found, constraint relaxation algorithms
[24] may be applied, or negotiations conducted between PL(j) planners, until a
solution is discovered. If no solution can be found, the planners report failure to
the job assignment submodule, and a new job assignment may be tried. If all
possible job assignments have been exhausted, the job assignment submodule
will report failure to the next higher level BG module.
   There is an executor EX(j) for each planner PL(j). The executor submodules
are responsible for successfully executing the plan state-graphs generated by
their respective planners. At each tick of the state clock, each executor measures
the difference between the current world state and its current plan subgoal state,
and issues a subcommand designed to null the difference. When the world
model indicates that a subtask in the current plan is successfully completed, the
executor steps to the next subtask in that plan. When all the subtasks in the
current plan are successfully executed, the executor steps to the first subtask in
the next plan. If the feedback indicates the failure of a planned subtask, the
executor branches immediately to a preplanned emergency subtask. Its planner
meanwhile begins work selecting or generating a new plan which can be
substituted for the former plan which failed. Output subcommands produced by

executors at level i become input commands to job assignment submodules in BG modules at level i-1.

3)   the executor sublevel -- EX(j) submodules

Planners PL(j) operate on the future. For each subsystem, there is a planner that is responsible for providing a plan that extends to the end of its planning horizon. Executors EX(j) operate in the present. For each subsystem, there is an executor that is responsible for monitoring the current (t=0) state of the world and executing the plan for its respective subsystem. Each executor performs a READ-COMPUTE-WRITE operation once each control cycle. At each level, each executor submodule closes a reflex arc, or servo loop. Thus, executor submodules at the various hierarchical levels form a set of nested servo loops. Executor loop bandwidths decrease on average about an order of magnitude at each higher level.

F.   The Task Decomposition Hierarchy

Task goals and task decomposition functions often have characteristic spatial and temporal requirements. There thus exists a functional hierarchy of task vocabularies that can be overlaid on the spatial/temporal hierarchy of Figure 4 [14]. For example:

Level 1 is where commands for coordinated positions, velocities, and forces of tools and end effectors are decomposed into drive signals to individual actuators. Sensory feedback is used to servo actuator positions, velocities, forces, and torques.
Level 2 is where commands for maneuvers of tools and end effectors are decomposed into smooth coordinated dynamically efficient trajectories. Sensory feedback servos coordinated trajectories of positions, velocities, forces, and torques.
Level 3 is where commands to manipulation, locomotion, and attention subsystems are decomposed into collision free paths that avoid obstacles and singularities. Sensory feedback servos movements relative to surfaces in the world.
Level 4 is where commands to an individual machine to perform simple tasks on single objects are decomposed into coordinated activity of locomotion, manipulation, attention, and communication subsystems. Sensory feedback signals task success or failure and sequences task activity.
Level 5 is where commands to closely coupled groups of machines, or workstations, are decomposed into coordinated behavior between individual machines. Sensory feedback signals group task success or failure and sequences interactions between each machine and nearby objects or persons.
Level 6 is where commands to manufacturing cells are decomposed into organized behavior of workstations. Sensory feedback signals success or failure of workstation tasks and signals location of batches of parts or tools.
Level 7 (arbitrarily the highest level) is where commands to the shop are decomposed into organized behavior of one or more cells. Sensory feedback indicates progress in filling orders.

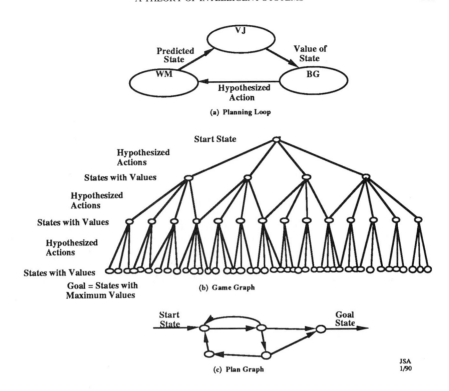

Figure 8.   The planning loop (a) produces a game graph (b). A trace in the game graph from the start state to a goal state is a plan that can be represented as a plan graph (c). Nodes in the game graph correspond to edges in the plan graph, and edges in the game graph correspond to nodes in the plan graph. Multiple edges exiting nodes in the plan graph correspond to conditional branches.

The mapping of BG functionality onto levels one through four defines the control functions necessary to control a single intelligent machine in performing simple task goals. Functionality at levels one through three is more or less fixed and specific to each type of machine. At level four and above, the mapping becomes more task and situation dependent. Levels five and above define the control functions necessary to control the relationships of an individual relative to others in groups, multiple groups, and the world as a whole.

Of course, not all manufacturing systems have the same number of levels. Some enterprises may have several shops that could be organized and controlled by a level 8 facility controller. Some machines may require higher speed servo loops that could be controlled by a level 0 controller. Many systems have discrete control functions that require no obstacle avoidance, dynamic, or servo computations. For these functions, levels 2 and 3 may be missing, or by-passed, and level 1 implemented by a simple relay or switch.

IV. The World Model

The world model is an intelligent system's internal representation of the external world. It is the system's best estimate of objective reality. It provides an interface between sensory processing and task decomposition. The world model is hierarchically organized so as to provide multiple levels of resolution in space and time.

Knowledge in the world model database includes both a-priori information which is available to the intelligent system before action begins, and a-posterior knowledge which is gained from sensing the environment as action proceeds. World model knowledge includes information about space, time, entities, events, and states of the world, including states of the system itself. The correctness and consistency of world model knowledge is verified by sensory processing mechanisms that measure differences between world model predictions and sensory observations.

WM modules provide memory, communication, and switching services that make the world model behave like a knowledge database in response to queries and updates from the BG, WM, SP, and VJ modules. The WM module in each node contains database management processes such as query processors, data servers, and question answering systems. The WM module also provides communication windows (the equivalent of a network terminal, or mailbox interface) into the knowledge database for each of the BG, WM, SP, and VJ modules in that node. Together the WM and KD modules make up the world model.

A. WM and KD Modules

The world model is hierarchically structured and distributed such that there is a WM and KD module in each node at every level of the control hierarchy. At each level, the WM modules perform the functions illustrated in Figure 9 and described below.

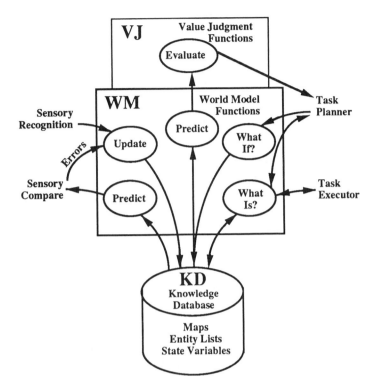

Figure 9. Functions performed by the WM module. 1) Update the knowledge database with recognized entities. 2) Predict sensory data. 3) Answer "What is?" queries from task executor and return current state of world. 4) Answer "What if?" queries from task planner and predict results for evaluation.

1) WM modules maintain the KD knowledge database, keeping it current and consistent. In this role, the WM modules perform the functions of a database management system. They update KD state estimates based on correlations and differences between world model predictions and sensory observations at each hierarchical level. The WM modules enter newly recognized entities, states, and events into the KD database, and delete entities and states determined by the sensory processing modules to no longer exist in the external world. The WM modules also enter estimates, generated by the VJ modules, of the reliability of KD state variables. Believability or confidence factors are assigned to many types of state variables.

2) WM modules generate predictions of expected sensory input for use by the appropriate sensory processing SP modules. In this role, a WM module performs the functions of a graphics engine, or state predictor, generating predictions that enable the sensory processing system to perform correlation and predictive filtering. WM predictions are based on the state of the task and estimated states of the external

world. For example in vision, a WM module may use the information in an object frame to generate predicted images which can be compared pixel by pixel, or entity by entity, with observed images.

3) WM modules answer "What is?" questions asked by the planners and executors in corresponding BG modules. In this role, the WM modules perform the function of database query processors, question answering systems, or data servers. World model estimates of the current state of the world are used by BG module planners as a starting point for planning. Current state estimates are also used by BG module executors for servoing and branching on conditions.

4) WM modules answer "What if?" questions asked by the planners in the corresponding level BG modules. In this role, the WM modules perform the function of simulation by generating expected status resulting from actions hypothesized by the BG planners. Results predicted by WM simulations are sent to value judgment VJ modules for evaluation. For each BG hypothesized action, a WM prediction is generated, and a VJ evaluation is returned to the BG planner. This BG-WM-VJ loop enables BG planners to select the sequence of hypothesized actions producing the best evaluation as the plan to be executed.

Data structures for representing explicit knowledge are defined to reside in a knowledge database that is hierarchically structured and distributed such that there is a knowledge database for each WM module in each node at every level of the system hierarchy. The communication system provides data transmission and switching services that make the WM modules and the knowledge database behave like a global virtual common memory in response to queries and updates from the BG, SP, and VJ modules. The communication interfaces with the WM modules in each node provides a window into the knowledge database for each of the computing modules in that node.

B.   Knowledge Representation

Knowledge in the world model database includes both a-priori information which is available to the intelligent system before action begins, and a-posterior knowledge which is gained from sensing the environment as action proceeds. World model knowledge includes information about space, time, entities, events, and states of the external world.

Knowledge about space is represented in maps. Knowledge about entities, events, and states is represented in lists and frames. Knowledge about the laws of physics, chemistry, optics, and the rules of logic and mathematics is represented in the WM functions that generate predictions and simulate results of hypothetical actions. Such knowledge may be represented as algorithms, or as IF/THEN rules of what happens under certain situations, such as when things are pushed, thrown, or dropped.

The world model also includes knowledge about the intelligent system itself, such as the state of all the currently executing processes in each of the BG, SP, WM, and VJ modules; the values assigned to goal priorities, attribute values assigned to objects, and events; parameters defining kinematic and dynamic models of robot arms or machine tool stages; state variables describing internal pressure, temperature, clocks, fuel levels, body fluid chemistry, etc.

The correctness and consistency of world model knowledge is verified by sensors and sensory processing SP mechanisms that measure differences between world model predictions and sensory observations.

C. Space

Gibson [25] has shown that the perception of space is primarily in terms of "medium, substance, and the surfaces that separate them." Medium is the air, water, fog, smoke, or falling snow through which the world is viewed. Substance is the material, such as earth, rock, wood, metal, flesh, grass, clouds, or water, that comprise the interior of objects. The surfaces that separate the viewing medium from the viewed objects is what are observed by the sensory system. The sensory input thus describes the external physical world primarily in terms of surfaces.

Surfaces are thus selected as the fundamental element for representing space in the proposed KD database. Volumes are treated as distances between surfaces. Objects are defined as circumscribed, often closed, surfaces. Lines, points and vertices lie on, and may define surfaces. Spatial relationships on surfaces are represented by maps.

D. Maps

A map is a two dimensional database that defines a mesh or grid on a surface. The surface represented by a map may be, but need not be, flat. For example, a map may be defined on a surface that is draped over, or even wrapped around, a 3-dimensional volume.

Maps can be used to describe the distribution of entities in space. It is always possible and often useful to project the physical 3-D world onto a 2-D surface defined by a map. For example, most commonly used maps are produced by projecting the world onto the 2-D surface of a flat sheet of paper, or the surface of a globe. One great advantage of such a projection is that it reduces the dimensionality of the world from three to two. This produces an enormous saving in the amount of memory required for a database representing space. The saving may be as much as three orders of magnitude, or more, depending on the resolution along the projected dimension.

E. Map Overlays

Most of the useful information lost in the projection from 3-D space to a 2-D surface can be preserved through the use of map overlays.

A map overlay is an assignment of values, or parameters, to points on the map. A map overlay can represent spatial relationships between 3-D objects. For example, an object overlay may indicate the presence of buildings, roads, bridges, and landmarks at various places on the map. Objects that appear smaller than a pixel on a map can be represented as icons. Larger objects may be represented by labeled regions that are projections of the 3-D objects on the 2- D map. Objects appearing on the map overlay may be cross referenced to an object database frame elsewhere in the world model. Information about the 3-D geometry of objects on the map may be represented in the object frame database.

Map overlays can also indicate attributes associated with points (or pixels) on the map. One of the most common map overlays defines terrain elevation. A value of terrain elevation (z) overlaid at each (x,y) point on a world map produces a topographic map.

A map can have any number of overlays. Map overlays may indicate brightness, color, temperature, even "behind" or "in-front." A brightness or color overlay may correspond to a visual image. For example, when aerial photos or satellite images are registered with map coordinates, they become brightness or color map overlays.

Map overlays may be useful for a variety of functions. In manufacturing, a mechanical scale drawing of a part or assembly is a map overlay. Drawings can indicate the projected shape of parts, the projected relative positions of parts in an assembly, surface finish, and geometrical dimensions and tolerances.

For mobile vehicles, map overlays may indicate terrain type, or region names, or can indicate values, such as cost or risk, associated with regions. Map overlays can indicate which points on the ground are visible from a given location in space. Overlays may also indicate contour lines, or grid lines such as latitude and longitude or range and bearing.

Map overlays showing object shape or terrain elevation may be useful for planning and executing tasks of manipulation and locomotion. Object and surface feature overlays can be useful for analyzing scenes and recognizing objects and places.

A map typically represents a snapshot of the world at a single instant in time. Motion can be represented by overlays of state variables such as velocity or image flow vectors, or traces (i.e. trajectories), of entity locations. Time may be represented explicitly by a numerical parameter associated with each trajectory point, or implicitly by causing trajectory points to fade, or be deleted, as time passes.

F. Map Pixel Frames

Definition: A map pixel frame is a frame that contains attributes and attribute-values attached to that map pixel.

Theorem: A set of map overlays are equivalent to a set of map pixel frames.

Proof: If each map overlay defines a parameter value for every map pixel, then the set of all overlay parameter values for each map pixel defines a frame for that pixel. Conversely, the frame for each pixel describes the region covered by that pixel. The set of all pixel frames thus defines a set of map overlays, one overlay for each attribute in the pixel frames. QED

For example, a pixel frame may describe the color, range, and orientation of the surface covered by the pixel. It may describe the name of (or pointer to) the entities to which the surface covered by the pixel belongs. It may also contain the location, or address, of the region covered by the pixel in other coordinate systems.

In the case of a video image, a map pixel frame might have the following form:

PIXEL_NAME =	location index on map (AZ, EL)
	(Sensor egosphere coordinates)
brightness	I
color	I_r I_b, I_g
spatial brightness gradient	dI/dAZ, dI/dEL (sensor egosphere coordinates)
temporal brightness gradient	dI/dt
image flow direction	B(velocity egosphere coordinates) [37]
image flow rate	dA/dt (velocity egosphere coordinates)
range	R to surface covered (from egosphere origin)
head egosphere location coord.)	az, el of egosphere ray to surface covered (head e.s.)
inertial egosphere location	a, e of egosphere ray to surface covered (inertial e.s.)
object map location	X, Y, Z of surface covered (object coordinates)
world map location	x, y, z of map point on surface covered (world

	coordinates)
linear feature pointer	pointer to frame of line, edge, or vertex covered by pixel
surface feature pointer	pointer to frame of surface covered by pixel
object pointer	pointer to frame of object covered by pixel
group pointer	pointer to group covered by pixel

Indirect addressing through pixel frame pointers can allow value state-variables assigned to objects or situations to be inherited by map pixels. For example, value state-variables such as attraction-repulsion, love-hate, fear-comfort assigned to objects and map regions can also be assigned through inheritance to individual map and egosphere pixels.

G. Map Resolution

The resolution for a world model map depends on how the map is generated and how it is used. For predicting sensory input, world model maps have resolution comparable to the resolution of the sensory system. For vision, the required map resolution may be on the order of a hundred thousand to a million pixels. For other sensory modalities, it may be considerably less.

For planning, different levels of the control hierarchy require maps of different scale. At higher levels, plans cover long distances and times, and require maps of large area, but low resolution. At lower levels, plans cover short distances and times, and maps cover small regions of space with high resolution.

In manufacturing, maps of different scales may be used to portray drawings of parts or assemblies, or the arrangement of machines, furniture, and workstations on the factory floor. Shop floor maps may be used for routing materials and tools between workstations. Workstation maps may be used for manipulator path planning. Assembly drawings may be used by processes at the workstation for planning assembly sequences. Part tray and fixture maps and part surface maps may be used for planning sequences of elementary moves for manipulation and machining. Part feature maps may be used for planning inspection routines.

The high resolution spatial memory of an intelligent creature typically consists of a finite number of relatively small regions that may be widely separated in space. For example, maps of workstations, parts, and part features may be represented in high resolution -- but part drawings depict only a tiny fraction of the space in a manufacturing environment. Workstations are typically connected by pathways that contain at most a few hundred known waypoints and branchpoints. The remainder of the world is known little, or not at all. Unknown regions, which make up the vast majority of the real world, occupy little or no space in the world model.

The efficient storage of maps with extremely non-uniform resolution can be accomplished in a variety of ways; for example, by quadtree [26], hash coding, or other sparse memory representations [29]. Pathways between known areas can be economically represented by graph structures (for example, highway maps). Sparse distributed memory [27] and neural net representations such as CMAC [28] suggest methods by which non-uniformly dense spatial information can be represented, in either computers or biological brains.

H. Maps and Egospheres

There are three general types of map coordinate frames that are important to an intelligent system: world coordinates, object coordinates, and egospheres.

1. World coordinates

World coordinates are often expressed in a Cartesian frame, and referenced to a point in the world. In most cases, the origin is an arbitrary point on the ground. The z axis is defined by the vertical, and the x and y axes define points on the horizon. For example, y may point North and x East. The value of z is often set to zero at sea level.

World coordinates may also be referenced to a moving point in the world. For example, the origin may be the self, or some moving object in the world. In this case, stationary pixels on the world map must be scrolled as the reference point moves.

An intelligent system may maintain several world maps with different resolutions and ranges.

2. Object Coordinates

Object coordinates are defined with respect to features in an object. For example, most engineering drawings of prismatic parts define the origin to be a vertex of the object shown. The coordinate axes are then defined by orthogonal edges or faces of the object. For turned objects, coordinate axes are typically defined by the center of rotation. For purposes of sensory processing, the origin of an object map might be defined as the center of gravity, with the coordinate axes defined by axes of symmetry, faces, edges, vertices, or skeletons [29]. There are a variety of surface representations that have been suggested for representing object geometry. Among these are generalized cylinders [30,31], B-splines [32], quadtrees [26], and aspect graphs [33].

3. Egospheres

An egosphere is a 2-dimensional spherical surface that is a map of the world as seen by an observer at the center of the sphere. Visible points on regions or objects in the world are projected on the egosphere wherever the line of sight from a sensor at the center of the egosphere to the points in the world intersect the surface of the sphere. Egosphere coordinates thus are polar coordinates defined by the self at the origin. As the self moves, the projection of the world flows across the surface of the egosphere.

Just as the world map is a flat 2-D (x,y) array with multiple overlays, so the egosphere is a spherical 2-D (AZ,EL) array with multiple overlays. Egosphere overlays can attribute brightness, color, range, image flow, texture, and other properties to regions and entities on the egosphere. Regions on the egosphere can thus be segmented by attributes, and egosphere points with the same attribute value may be connected to form contour lines. Egosphere overlays may also indicate the trace, or history, of brightness values or entity positions over some time interval. Objects may be represented on the egosphere by icons, and each object may have in its database frame a trace, or trajectory, of positions on the egosphere over some time interval.

4. Map Transformations

If surfaces in real world space are covered by an array (or map) of points in a coordinate system defined in the world, and the surface of the WM egosphere is also represented as an array of points, then there exists a function G that transforms each point on the real world map into a point on the WM egosphere, and a function G' that transforms each point on the WM egosphere into a point on the real world map.

For example, Figure 10 shows the 3-D relationship between an egosphere and world map coordinates. For every point (x,y,z) in world coordinates, there is a point (AZ,EL,R) in ego centered coordinates which can be computed by the 3x3 matrix function G

$$(AZ,EL,R)^T = G\ (x,y,z)^T$$

There, of course, may be more than one point in the world map that gives the same (AZ,EL) values on the egosphere. Only the (AZ,EL) with the smallest value of R will be visible to an observer at the center of the egosphere. The deletion of egosphere pixels with R larger than the smallest for each value of (AZ,EL) corresponds to the hidden surface removal problem common in computer graphics.

For each egosphere pixel where R is known, an (x,y,z) can be computed from (AZ,EL,R) by the function G'

$$(x,y,z)^T = G'\ (AZ,EL,R)^T$$

Any point in the world topological map can thus be projected onto the egosphere (and vice versa when R is known). Projections from the egosphere to the world map will leave blank those map pixels that cannot be observed from the center of the egosphere.

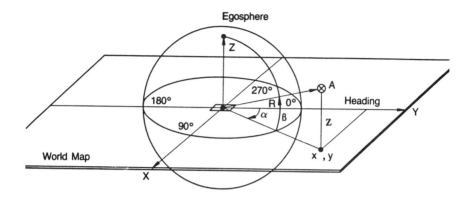

Figure 10.   Geometric relationship between world map and egosphere coordinates.

As ego motion occurs (i.e. as the self object moves through the world), the egosphere moves relative to world coordinates, and points on the egocentric maps flow across their surfaces.   Ego motion may involve translation, or rotation, or both; in a stationary world, or a world containing moving objects. Once range to any stationary point in the world is known, its pixel motion on the egosphere can be predicted from knowledge of egomotion. For moving points, prediction of pixel motion on the egosphere requires additional knowledge of object motion.

Transformations to and from the egosphere and the world map allow the intelligent system to sense the world from one perspective and interpret it in another. They allow the intelligent system to compute how entities in the world would look from another viewpoint. They provide the ability to overlay sensory input with world model predictions, and to compute the geometrical and dynamical functions necessary to navigate, focus attention, and direct action relative to entities and regions of the world.

All of the above egosphere transformations can be inverted, so that conversions can be made in either direction. Each transformation consists of a relatively simple vector function that can be computed for each pixel in parallel. Full image egosphere transformations can be accomplished at television frame rates by state-of-the-art serial computing hardware. They can be accomplished in microseconds by parallel hardware.

I. Entities

Definition:

An entity is an element from the set {point, line, surface, object, group, group^2, . . .}

The world model contains information about entities stored in lists, or frames. The KD database contains a list of all the entities that the intelligent system knows about. A subset of this list is the set of current entities known to be present in any given situation. A subset of the list of current entities is the set of entities of attention.

There are two types of entities: generic and specific. A generic entity is an example of a class of entities. A generic entity frame contains the attributes of its class. A specific entity is a particular instance of an entity. A specific entity frame inherits the attributes of the class to which it belongs.

An example of an entity frame is:

ENTITY NAME	--	name of entity
kind	--	name of class, species, or model of entity
type	--	generic or specific point, line, surface, object, or group
position	--	world map coordinates (uncertainty)
		egosphere coordinates (uncertainty)
dynamics	--	velocity (uncertainty)
		acceleration (uncertainty)
trajectory	--	sequence of positions
geometry	--	center of gravity (uncertainty)
		axis of symmetry (uncertainty)
		size (uncertainty)
		boundaries (uncertainty)
links	--	subentities
		parent entity
properties	--	physical
		mass
		color
		substance
		behavioral
		individual
		social (of animate objects)
capabilities	--	speed, range
value state-variables	--	attract-repulse
		confidence-fear
		love-hate

For example, an entity frame of a typical mechanical part to be manufactured might have the following values:

ENTITY NAME	--	part serial #
kind	--	model #, or group technology code
type	--	desired object
position	--	x,y,z (in tray map coordinates)
	--	AZ, EL, R (in egosphere image of robot camera)

dynamics	--	velocity = 0
		acceleration = 0
trajectory	--	sequence of positions over past 10 seconds
geometry	--	prismatic
		height, width, depth
		volume
		boundaries
topology	--	relationships between subentities
links	--	subentities - surfaces (front, top, bottom, side, holes,
grooves,		etc.)
	--	parent entity - group (tray #)
properties	--	physical
		mass (35 lbs)
		reflectance (shiny)
		substance (brass)
value state-variables	--	sunk cost
	--	cost to complete
	--	priority

At any time there exists a set of entities that the system has in its knowledge base. Of this set, there is a subset of entities that are currently in the workspace. Of this subset, there is a subset of entities of attention. These are the entities that currently are visible to the sensor egosphere, or are objects of currently active tasks.

J. Map - Entity Relationship

Map and entity representations can be cross referenced and tightly coupled by real-time computing hardware. Each pixel on the map can have in its frame a pointer to the list of entities covered by that pixel. For example, each pixel may cover a point entity indicating brightness, color, spatial and temporal gradients of brightness and color at a point. Each pixel may also cover a linear entity indicating an edge, image flow, or range; a surface entity indicating slope, boundaries, and range discontinuities; an object entity indicating the name of the object covered; a group entity indicating the name of the group covered, etc.

Likewise, each entity has in its frame a set of position and orientation variables that enables the geometry engine to compute the set of egosphere or world map pixels covered by each entity. Thus, entity parameters associated with objects, surfaces, edges, and vertices can be overlaid on the world and egosphere maps.

Cross referencing between pixel maps and entity frames occurs at every level of the system hierarchy. Each level of sensory processing computes entity attributes and adds overlays to the map representations. The entity database can be updated by the recognition of image parameters at points on the egosphere, and the map database can be predicted from knowledge of entity parameters in the world model. Many of the attributes in entity frames are time dependent state-variables. Each time dependent state-variable may possess a short term memory queue wherein is stored a state trajectory, or trace, that describes its recent temporal history.

At each hierarchical level, temporal traces stretch backward about as far as the planning horizon at that level stretches into the future. At each level, the historical trace of an entity state-variable may be summarized, or integrated, into a single event in the historical trace of the next higher level. Entity state-variable histories may be summarized by running averages, Fourier transform coefficients, Kalman filter parameters, or other recursive state-estimation techniques.

Each state-variable in an entity frame may have value state-variable parameters that indicate levels of believability, confidence, support, or plausibility, and measures of dimensional uncertainty. These are computed at each level by value judgment functions that reside in corresponding VJ modules.

Value state-variable parameters may be overlaid on the map and egosphere regions where the entities to which they are assigned appear. This facilitates planning. For example, approach-avoidance behavior can be planned on an egosphere map overlay defined by the summation of attractor and repulsor value state-variables assigned to objects or regions that appear on the egosphere. Vehicle navigation planning can be done on a map overlay whereon risk and benefit values are assigned to regions on the egosphere or world map.

K. ENTITY DATABASE HIERARCHY

The entity database is hierarchically structured. Each entity consists of a set of subentities, and is part of a parent entity. For example, an object may consist of a set of surfaces, and be part of a group.

The definition of an object is quite arbitrary, however, at least from the point of view of the world model. For example, is a nose an object? If so, what is a face? Is a head an object? Or is it part of a group of objects comprising a body?

Only in the context of a task does the definition of an object become clear. For example, in a task frame, an object may be defined either as the agent, or as acted upon by the agent executing the task. Thus, in the context of a specific task, the nose (or face, or hand) may become an object because it appears in a task frame as the agent or object of a task.

Thus, perception in an intelligent system is task (or goal) driven, and the structure of the world model entity database is defined by, and may be reconfigured by, the nature of goals and tasks. It is, therefore, not necessarily the role of the world model to define the boundaries of entities, but rather to represent the boundaries defined by the task frame, and to map regions and entities circumscribed by those boundaries with sufficient resolution to accomplish the task. It is the role of the sensory processing system to identify regions and entities in the external real world that correspond to those represented in the world model, and to discover boundaries that circumscribe objects defined by tasks.

This suggests that the hierarchical structure of the world model map and entity database might be placed in one-to-one correspondence with the hierarchical levels of task decomposition proposed in Section 3. For example:

At level 1 of the task decomposition hierarchy, the world model can represent map overlays and frames for point entities in head egosphere coordinates. In the case of vision, point entities may consist of brightness or color intensities, spatial and temporal derivatives of those intensities, image flow vectors, and range estimates for each pixel. These representations are roughly analogous to Marr's "primal sketch" [34], and are compatible with experimentally observed data representations in the primary visual cortex (V1) [35].

At level 2, the world model can represent map overlays and frames for linear entities consisting of clusters, or strings, of point entities. In the visual system, linear entities may consist of connected edges (brightness, color, or depth), vertices, and trajectories of points in space/time. Attributes such as 3-D position, orientation, velocity, and rotation are represented in the frame of each linear entity. These representations are compatible with experimentally observed data representations in the secondary visual cortex (V2) [36].

At level 3, the world model can represent map overlays for surface entities computed from sets of linear entities clustered or swept into bounded surfaces or maps, such as terrain maps, B-spline surfaces, or general functions of two variables.

Surface entity frames contain transform parameters to and from object coordinates. In the case of vision, entity attributes may describe surface color, texture, surface position and orientation, velocity, size, rate of growth in size, shape, and surface discontinuities or boundaries. Level 3 is thus roughly analogous to Marr's "2 1/2-D sketch," and is compatible with known representation of data in visual cortical areas V3 and V4.

At level 4, the world model can represent map overlays for object entities computed from sets of surfaces clustered or swept so as to define 3-D volumes, or objects. Object entity frames contain transform parameters to and from object coordinates. Object entity frames may also represent object type, position, translation, rotation, geometrical dimensions, surface properties, occluding objects, contours, axes of symmetry, volumes, etc. These are analogous to Marr's "3-D model" representation, and compatible with data representations in visual association areas of parietal and temporal cortex.

At level 5, the world model can represent map overlays for group entities consisting of sets of objects clustered into groups or packs. Group entity frames contain transform parameters to and from world coordinates. Group entity frames may also represent group species, center of mass, density, motion, map position, geometrical dimensions, shape, spatial axes of symmetry, volumes, etc.

At level 6, the world model can represent map overlays for sets of group entities clustered into groups of groups, or group2 entities. At level 7, the world model can represent map overlays for sets of group2 entities clustered into group3 (or world) entities, and so on. At each higher level, world map resolution decreases and range increases by about an order of magnitude per level.

The highest level entity in the world model is the world itself, i.e. the environment as a whole. The environment entity frame contains attribute state-variables that describe the state of the environment, such as temperature, wind, precipitation, illumination, visibility, etc.

L. Events

An event is a state, condition, or situation that exists at a point in time, or occurs over an interval in time. Events may be represented in the world model by frames with attributes such as the point, or interval, in time and space when the event occurred, or is expected to occur. Event frame attributes may indicate start and end time, duration, type, relationship to other events, etc.

An example of an event frame is:

EVENT NAME	--	event identifier
kind	--	class or species
type	--	generic or specific
modality	--	visual, auditory, tactile, etc.
time	--	when event detected
interval	--	period over which event took place
position	--	map location where event occurred
links	--	parent event subevents
value	--	good-bad, benefit-cost, etc.

State-variables in the event frame may have confidence levels, degrees of support and plausibility, and measures of dimensional uncertainty similar to those in spatial entity frames. Confidence state-variables may indicate the degree of certainty that an event actually occurred, or was correctly recognized.

The event frame database is hierarchical. At each level of the sensory processing hierarchy, the recognition of a pattern, or string, of level(i) events makes up a single level(i+1) event. We can thus place the hierarchical levels of the event frame database in one-to-one correspondence with the hierarchical levels of task decomposition and sensory processing. For example at:

Level 1 -- an event may span a few milliseconds. A typical level one event might be the measurement of a joint position, force, or velocity, the recognition of a tone, hiss, click, a change in pixel intensity, or a measurement of intensity gradient or image flow at a pixel.

Level 2 -- an event may span a few tenths of a second. A typical level two event might be the recognition of a tool path, or trajectory of a visual point or feature.

Level 3 -- an event may span a few seconds, and consist of the recognition of a series of tool path segments, or the motion of a visual surface.

Level 4 -- an event may span a few tens of seconds, and consist of the recognition of a robot or machine tool task, or a visual observation of object motion.

Level 5 -- an event may span a few minutes and consist of recognition of a workstation task.

Level 6 -- an event may span an hour and consist of recognition of a series of a cell task.

Level 7 -- an event may span a day and consist of observations of an entire day's activities in a shop.

State-variables in the event frame may have confidence levels, degrees of support and plausibility, and measures of dimensional uncertainty similar to those in spatial entity frames. Additional confidence state-variables may indicate the degree of certainty that an event actually occurred, or was correctly recognized.

IV. SENSORY PROCESSING

Sensory processing is the mechanism of perception. Perception is the establishment and maintenance of correspondence between the internal world model and the external real world. The function of sensory processing is to extract information about entities, events, states, and relationships in the external world, so as keep the world model accurate and up to date.

A. Measurement of Surfaces

World model maps are updated by sensory measurement of points, edges, and surfaces. Such information is usually derived from vision or touch sensors, although some intelligent systems may derive it from sonar, radar, or laser sensors.

The most direct method of measuring points, edges, and surfaces is through touch. Many creatures, from insects to mammals, have antennae or whiskers that are used to measure the position of points and orientation of surfaces in the environment. Virtually all creatures have tactile sensors in the skin, particularly in the digits, lips, and tongue. Proprioceptive sensors indicate the position of the feeler or tactile sensor relative to the self when contact is made with an external surface. This, combined with knowledge of the kinematic position of the feeler endpoint, provides the information necessary to compute the position on the egosphere of each point contacted. A series of felt points defines edges and surfaces on the egosphere.

In the manufacturing environment, measurements made by touch probes are used by coordinate measuring machines to compute the position of points. Touch probes on machine tools and tactile sensors on robots can be used for making similar measurements. From such data, sensory processing algorithms can compute the orientation and position of surfaces, the shape of holes, the distance between surfaces, and the dimensions of parts.

Other methods for measuring surfaces include stereo vision, photogrammetry, laser ranging systems, structured light, image flow, acoustic ranging systems, and focus-based optical probes.   Each of these various methods produce different accuracies and have differing computational and operational requirements.

For example, stereo vision and photogrammetry require two cameras separated by a known distance, and a method, such as cross correlation, for computing the angular disparity between corresponding points in images from the two cameras. Accuracy depends on camera separation and resolution.   Image flow requires knowledge of relative motion between a single camera and the viewed object, plus a method for computing temporal and spatial derivatives of brightness along flow lines [37,38]. Accuracy depends on camera resolution and image sample rates. Stereo is useful for manipulation of 3-D objects. Image flow is useful for mobility and obstacle avoidance in a complex environment. Laser ranging and structured light can be used for either manipulation or mobility, and require less computational power than stereo or image flow methods. However, they are dependent on surface reflectance and typically require longer to acquire an image. Acoustic ranging is simple and inexpensive, but notoriously inaccurate and unreliable. Focus-based optical probes typically operate only within a few millimeters of a surface, but they can be almost as accurate as touch probes and are considerably faster because they do not require contact.

All of the above methods for deriving surfaces are primitive in the sense that they compute directly from sensory input without recognizing entities or understanding anything about the scene. Depth measurements from primitive processes can immediately generate maps that can be used directly by the lower levels of the behavior generating hierarchy to avoid obstacles and approach surfaces.

Additional information about surface position and orientation may also be computed from shading, shadows, and texture gradients. These methods typically depend on higher levels of visual perception such as geometric reasoning, recognition of objects, detection of events and states, and the understanding of scenes.

B.   Recognition and Detection

Recognition is the establishment of a one-to-one match, or correspondence, between a real world entity and a world model entity .

The process of recognition may proceed top-down, or bottom-up, or both simultaneously. For each entity in the world model, there exists a frame filled with information that can be used to predict attributes of corresponding entities observed in the world. The top-down process of recognition begins by hypothesizing a world model entity and comparing its predicted attributes with those of the observed entity. When the similarities and differences between predictions from the world model and observations from sensory processing are integrated over a space-time window that covers an entity, a matching, or cross-correlation value is computed between the entity and the model. If the correlation value rises above a selected threshold, the entity is said to be recognized. If not, the hypothesized entity is rejected and another tried

The bottom-up process of recognition consists of applying filters and masks to incoming sensory data, and computing image properties and attributes. These may then be stored in the world model, or compared with the properties and attributes of entities already in the world model. Both top-down and bottom-up processes proceed until a match is found, or the list of world model entities is exhausted. Many perceptual matching processes may operate in parallel at multiple hierarchical levels simultaneously.

If a SP module recognizes a specific entity, the WM at that level updates the attributes in the frame of that specific WM entity with information from the sensory system.

If the SP module fails to recognize a specific entity, but instead achieves a match between the sensory input and a generic world model entity, a new specific WM entity will be created with a frame that initially inherits the features of the generic entity. Slots in the specific entity frame can then be updated with information from the sensory input.

If the SP module fails to recognize either a specific or a generic entity, the WM may create an "unidentified" entity with an empty frame. This may then be filled with information gathered from the sensory input.

When an unidentified entity occurs in the world model, the behavior generation system may (depending on other priorities) select a new goal to <identify the unidentified entity>. This may initiate an exploration task that positions and focuses the sensor systems on the unidentified entity, and possibly even probes and manipulates it, until a world model frame is constructed that adequately describes the entity. The sophistication and complexity of the exploration task depends on task knowledge about exploring things. Such knowledge may be very advanced and include sophisticated tools and procedures, or very primitive. Entities may, of course, simply remain labeled as "unidentified," or "unexplained."

Detection of events is analogous to recognition of entities. Observed states of the real world are compared with states predicted by the world model. Similarities and differences are integrated over an event space-time window, and a matching, or cross-correlation value is computed between the observed event and the model event. When the cross-correlation value rises above a given threshold, the event is detected.

C. The Context of Perception

If, as suggested in Figure 4, there exists in the world model at every hierarchical level a short term memory in which is stored a temporal history consisting of a series of past values of time dependent entity and event attributes and states, it can be assumed that at any point in time, an intelligent system has a record in its short term memory of how it reached its current state. Figures 4 and 5 also imply that, for every planner in each behavior generating BG module at each level, there exists a plan, and that each executor is currently executing the first step in its respective plan. Finally, it can be assumed that the knowledge in all these plans and temporal histories, and all the task, entity, and event frames referenced by them, is available in the world model.

Thus it can be assumed that an intelligent system almost always knows where it is on a world map, knows how it got there, where it is going, what it is doing, and has a current list of entities of attention, each of which has a frame of attributes (or state variables) that describe the recent past, and provide a basis for predicting future states. This includes a prediction of what objects will be visible, where and how object surfaces will appear, and which surface boundaries, vertices, and points will be observed in the image produced by the sensor system. It also means that the position and motion of the eyes, ears, and tactile sensors relative to surfaces and objects in the world are known, and this knowledge is available to be used by the sensory processing system for constructing maps and overlays, recognizing entities, and detecting events.

Were the above not the case, the intelligent system would exist in a situation analogous to a person who suddenly awakens at an unknown point in space and time. In such cases, it typically is necessary even for humans to perform a series of tasks designed to "regain their bearings," i.e. to bring their world model into correspondence with the state of the external world, and to initialize plans, entity frames, and system state variables.

It is, of course, possible for an intelligent system to function in a totally unknown environment, but not well, and not for long. Not well, because no system can behave intelligently without making use of historical information that forms the context of its current task. Without information about where it is, and what is going on, even the most intelligent creature is lost and confused. Not for long, because the sensory processing system continuously updates the world model with new information about the current situation and its recent historical development, so that, within a few seconds, an intelligent system can usually acquire a functionally usable map and a usable set of entity state variables from the immediately surrounding environment.

D. Sensory Processing SP Modules

At each level of the proposed architecture, there are a number of computational nodes. Each of these contains an SP module, and each SP module consists of four sublevels, as shown in Figure 11.

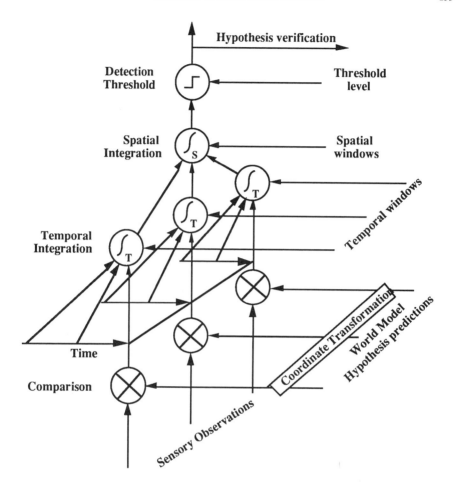

Figure 11. Each sensory processing SP module consists of: 1) a set of coordinate transformers, 2) a set of comparators that compare sensory observations with world model predictions, 3) a set of temporal integrators that integrate similarities and differences, 4) a set of spatial integrators that fuse information from different sensory data streams, and 5) a set of threshold detectors that recognize entities and detect events.

Sublevel 1 -- Comparison

Each comparison submodule matches an observed sensory variable with a world model prediction of that variable. This comparison typically involves arithmetic operations, such as multiplication or subtraction (and possibly scrolling and integration functions) which yield a measure of similarity (correlation) and difference (variance) between an observed variable and a predicted variable. Similarities indicate the degree to which the WM predictions are correct, and hence are a measure of the correspondence between the world model and reality.

Differences indicate a lack of correspondence between world model predictions and sensory observations. Differences imply that either the sensor data or world model is incorrect. Difference images from the comparator go three places:

a) They are returned directly to the WM for real-time local pixel and entity attribute updates. This produces a tight feedback loop whereby the world model predicted image becomes an array of recursive state-estimations. Difference images are thus error signals used to make each pixel and entity attribute of the predicted image a "best estimate" predictor of the corresponding pixel and entity attribute in the current sensory input.

b) They are also transmitted upward to the integration sublevels where they are integrated over time and space in order to recognize and detect global entity attributes. This integration constitutes a summation, correlation, or "chunking", of sensory data into entities. At each level, lower order entities are "chunked" into higher order entities, i.e. points are chunked into lines, lines into surfaces, surfaces into objects, objects into groups, etc.

c) They are transmitted to the VJ module at the same level where statistical parameters are computed in order to assign confidence and believability factors to pixel entity attribute estimates.

Sublevel 2 -- Temporal integration

Temporal integration submodules integrate similarities and differences between predictions and observations over intervals of time. Temporal integration submodules operating just on sensory data can produce a summary, such as a total, or average, of sensory information over a given time window. Temporal integrator submodules operating on the similarity and difference values computed by comparison submodules may produce temporal cross-correlation and covariance functions between the model and the observed data. These correlation and covariance functions are measures of how well the dynamic properties of the world model entity match those of the real world entity. The boundaries of the temporal integration window may be derived from world model prediction of event durations, or from behavior generation parameters such as sensor fixation periods.

Sublevel 3 -- Spatial integration

Spatial integrator submodules integrate similarities and differences between predictions and observations over regions of space. This produces spatial cross-correlation or convolution functions between the model and the observed data. Spatial integration summarizes sensory information from multiple sources at a single point in time. It determines whether the geometric properties of a world model entity match those of a real world entity. For example, the product of an edge operator and an input image may be integrated over the area of the operator to obtain the correlation between the image and the edge operator at a point. The limits of the spatial integration window may be determined by world model predictions of entity size. In some cases, the order of temporal and spatial integration may be reversed, or interleaved.

Sublevel 4 -- Recognition/Detection threshold

When the spatio-temporal correlation function exceeds some threshold, object recognition (or event detection) occurs. For example, if the spatio-temporal summation over the area of an edge operator exceeds threshold, an edge is said to be detected at the center of the area.

Figure 12 illustrates the nature of the SP-WM interactions between an intelligent vision system and the world model at two levels. On the left of Figure 12, the world of reality is viewed through the window of an egosphere such as exists in the primary visual cortex. On the right is a world model consisting of: 1) symbolic entity frames in which entity attributes are stored, and 2) an iconic predicted image that is registered in real-time with the observed sensory image. In the center of Figure 12, is a comparator where the expected image is subtracted from (or otherwise compared with) the observed image.

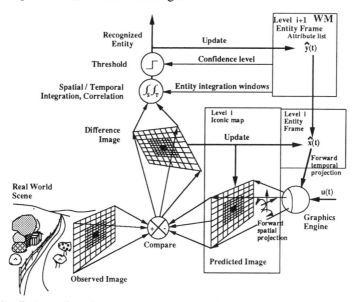

Figure 12. Interaction between world model and sensory processing. Difference images are generated by comparing predicted images with observed images. Feedback of differences produces a Kalman best estimate for each data variable in the world model. Spatial and temporal integration produce cross-correlation functions between the estimated attributes in the world model and the real world attributes measured in the observed image. When the correlation exceeds threshold, entity recognition occurs.

The level(i) predicted image is initialized by the equivalent of a graphics engine operating on symbolic data from frames of entities hypothesized at level(i+1) and level(i). The predicted image and the level(i) entity frames are updated by differences between the predicted image and the observed sensory input. By this process, the predicted image becomes the world model's "best estimate prediction" of the incoming sensory image, and a high speed loop is closed between the WM and SP modules at level(i).

When recognition occurs in level (i), the world model level(i+1) hypothesis is confirmed and both level(i) and level(i+1) symbolic parameters that produced the match are updated in the symbolic database. This closes a slower, more global, loop between WM and SP modules through the symbolic entity frames of the level(i+1) world model.

Many examples of these kinds of looping interaction can be found in the model matching and model-based recognition literature [39]. Similar closed loop filtering concepts have been used for years for signal detection, and for dynamic systems modeling in aircraft flight control systems. Recently they have been applied to high speed visually guided driving of autonomous ground vehicles [40].

The behavioral performance of intelligent biological creatures suggests that mechanisms similar to those shown in Figures 11 and 12 exist in the brain. In biological or neural network implementations, SP modules may contain thousands, even millions, of comparison submodules, temporal and spatial integrators, and threshold submodules. The neuroanatomy of the mammalian visual system suggests how maps with many different overlays, as well as lists of symbolic attributes, could be processed in parallel in real-time. In such structures it is possible for multiple world model hypotheses to be compared with sensory observations at multiple hierarchical levels, all simultaneously.

In artificially intelligent systems, the design of comparable SP systems remains a research topic. Certainly, most current machine vision systems are much simpler and less capable than what is suggested here. In manufacturing, perhaps the most advanced processing of sensory data is found in coordinate measuring machines where objects, surfaces, features, lines, and points are defined in "as-designed" and "as-measured" geometrical databases. In some advanced inspection systems the "as-designed" database can be used to define path plans for measurement probes, and "as-measured" dimensions can be compared with "as-designed" dimensions to determine whether tolerance specifications are met.

E. World Model Update

Attributes in the world model predicted image may be updated by a recursive estimation formula of the form

$$\hat{x}(t+1) = \hat{x}(t) + A \, \hat{y}(t) + B \, u(t) + K(t) \, [x(t) - \hat{x}(t)]$$

where
$\hat{x}(t)$ is the best estimate vector of world model i-order entity attributes at time t

A is a matrix that computes the expected rate of change of $\hat{x}(t)$ given the current best estimate of the i+1 order entity attribute vector $\hat{y}(t)$

B is a matrix that computes the expected rate of change of $\hat{x}(t)$ due to external input u(t)

K(t) is a confidence factor vector for updating $\hat{x}(t)$
The value of K(t) may be computed by a formula of the form

$$K(t) = Ks(j,t) \, [1 - Km(j,t)]$$

where
Ks(j,t) is the confidence in the sensory observation of the j-th real world attribute x(j,t) at time t $0 < Ks(j,t) < 1$
$K_m(j,t)$ is the confidence in the world model prediction of the j-th

attribute $\hat{x}$ (j,t) at time t $0 < K_m$ (j,t) < 1

The confidence factors (K_m and K_s) in the formula may depend on the statistics of the correspondence between the world model entity and the real world entity (e.g. the number of data samples, the mean and variance of $[x(t) - \hat{x}(t)]$, etc.). A high degree of correlation between x(t) and $\hat{x}(t)$ in both temporal and spatial domains indicates that entities or events have been correctly recognized, and states and attributes of entities and events in the world model correspond to those in the real world environment. World model data elements that match observed sensory data elements are reinforced by increasing the confidence, or believability factor, Km(j,t) for the entity or state at location j in the world model attribute lists. World model entities and states that fail to match sensory observations have their confidence factors Km(j,t) reduced. The confidence factor Ks(j,t) may be derived from the signal-to-noise ratio of the j-th sensory data stream.

The numerical value of the confidence factors may be computed by a variety of statistical methods such as Baysian or Dempster-Shafer statistics.

F. The Mechanisms of Attention

In order for an intelligent system to make best use of its sensing and computing resources, sensing and sensory processing must be active processes that are directed by goals and priorities generated in the behavior generating system.

In each node of the intelligent system hierarchy, the behavior generating BG modules should request specific information needed for the current task from sensory processing SP modules. By means of such requests, the BG modules control the processing of sensory information and focus the attention of the WM and SP modules on the entities and regions of space that are important to success in achieving behavioral goals. Requests by BG modules for specific types of information cause SP modules to select particular sensory processing masks and filters to apply to the incoming sensory data. Requests from BG modules enable the WM to select which world model data to use for predictions, and which prediction algorithm to apply to the world model data. BG requests may also define which correlation and differencing operators to use, and which spatial and temporal integration windows and detection thresholds to apply.

As shown in Figure 3, behavior generating BG modules in the attention subsystem should also actively point the eyes and ears, and direct touch probes and tactile sensors toward entities of attention. BG modules in the vision subsystem control the motion of cameras, adjust the iris and focus, and actively point the fovea (if one exists) to probe the environment for the visual information needed to pursue behavioral goals [41,42]. Similarly, BG modules in an acoustic subsystem may actively direct acoustic sensors and tune filters to mask background noises and discriminate in favor of the acoustic signals of importance to behavioral goals.

Because of the active nature of the attention subsystem, sensor resolution and sensitivity need not be uniformly distributed, but can be highly focused. For example, in primate vision systems, receptive fields of optic nerve fibers from the eye are several thousand times more densely packed in the fovea than near the periphery of the visual field. Receptive fields of biological touch sensors are also several thousand times more densely packed in the finger tips and on the lips and tongue, than on other parts of the body such as the torso.

The active control of sensors with non-uniform resolution has profound impact on the communication bandwidth, computing power, and memory capacity required by the sensory processing system. For example, there are roughly 500,000 fibers in the the optic nerve from a single human eye. These fibers are distributed such that about 100,000 are concentrated in the ±1.0 degree foveal region with resolution of about 0.007 degrees. About 100,000 cover the surrounding ± 3 degree region with resolution of about 0.02 degrees. 100,000 more cover the surrounding ± 10 degree region with resolution of 0.07 degrees. 100,000 more cover the surrounding ± 30 degree region with a resolution of about 0.2 degrees. 100,000 more cover the remaining ± 80 degree region with resolution of about 0.7 degree [43]. The total number of pixels is thus about 500,000 pixels, or somewhat less than that contained in two standard commercial TV images. Without non-uniform resolution, covering the entire visual field with the resolution of the fovea would require the number of pixels in about 6000 standard TV images. Thus, for a vision sensory processing system with any given computing capacity, active control and non-uniform resolution in the retina can produce more than three orders of magnitude improvement in image processing capability.

SP modules in the attention subsystem process data from low-resolution wide-angle sensors to detect regions of interest, such as entities that move, or regions that have discontinuities (edges and lines), or have high curvature (corners and intersections). The attention BG modules then actively maneuver the eyes, fingers, and mouth so as to bring the high resolution portions of the sensory systems to bear precisely on these points of attention. The result gives the subjective effect of high resolution everywhere in the sensory field. For example, wherever the eye looks, it sees with high resolution, for the fovea is always centered on the item of current interest.

The act of perception thus involves both sequential and parallel operations. For example, the fovea of the eye is typically scanned sequentially over points of attention in the visual field [41]. Touch sensors in the fingers are actively scanned over surfaces of objects, and the ears may be pointed toward sources of sound. While this sequential scanning is going on, parallel recognition processes hypothesize and compare entities at all levels simultaneously.

G.   The Sensory Processing Hierarchy

It has long been recognized that sensory processing occurs in a hierarchy of processing modules, and that perception proceeds by "chunking," i.e. by recognizing patterns, groups, strings, or clusters of points at one level as a single feature, or point in a higher level, more abstract space. It also has been observed that this chunking process proceeds by about an order of magnitude per level, both spatially and temporally [17,18]. Thus, at each level in the proposed architecture, SP modules integrate, or chunk, information over space and time by about an order of magnitude.

Figure 13 describes the nature of the interactions hypothesized to take place between the sensory processing and world modeling modules at the first four levels, as the recognition process proceeds. The functional properties of the SP modules are coupled to, and determined by, the predictions of the WM modules in their respective processing nodes. The WM predictions are, in turn, effected by states of the BG modules.

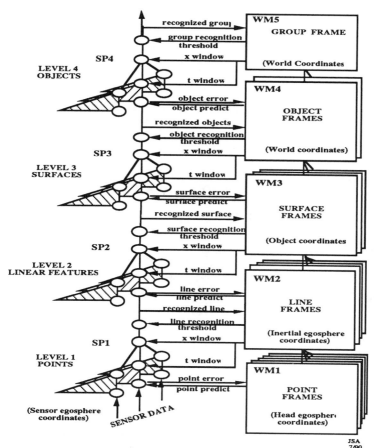

Figure 13. The nature of the interactions that take place between the world model and sensory processing modules. At each level, predicted entities are compared with observed. Errors are returned directly to the world model to update the model, as well as forwarded upward to be integrated over time and space windows provided by the world model. Correlations that exceed threshold are recognized as entities.

It has been further hypothesized that the sensory processing and world modeling modules shown in Figure 13 correspond closely to the hierarchical levels known to exist in the primate vision system [44].

Figure 14 illustrates the concept stated in this hypothesis. Visual input to the camera consists of photometric brightness and color intensities measured by photoreceptors. Brightness intensities are denoted by $I(k, AZ, EL, t)$, where I is the brightness intensity measured at time t by the pixel at sensor egosphere azimuth AZ and elevation EL of camera k. Intensity signals I may vary over time intervals on the order of a millisecond.

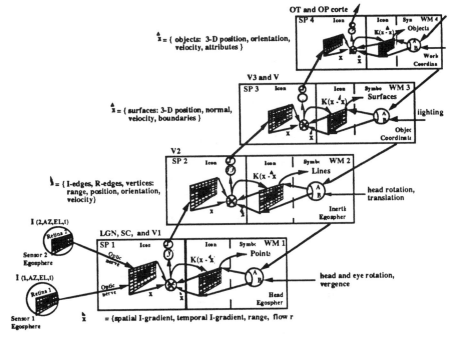

Figure 14. Hypothesized correspondence between levels in the proposed model and neuranatomical structures in the mammalian vision system. At each level, the WM module contains both iconic and symbolic representations. At each level, the SP module compares the observed image with a predicted image. At each level, both iconic and symbolic world models are updated, and map overlays are computed. LGN = lateral geniculate nuclei, OT - occipital-temporal, OP = occipital-parietal, SC = superior colliculus

Image preprocessing corresponding to that performed on the retina detects both spatial and temporal derivatives at each point in the visual field. Output from this preprocessing stage becomes input to sensory processing level 1 in Figure 14. Level 1 inputs correspond to events of a few milliseconds duration.

Level 1 inputs from two or more cameras can be overlaid such that the visual fields are in registration. Data from camera pointing sensors provides information about eye convergence, and pan, tilt, and roll of the camera relative to the sensor platform. This allows image map points in camera coordinates to be transformed into image map points in sensor platform coordinates (or vice versa) so that visual and tactile position data can be registered and fused. Registration of corresponding pixels from two separate cameras also provides the basis for range from stereo to be computed for each pixel [38]. Level 1 also computes the orientation of edges and lines at each pixel.

Level 1 of the sensory processing hierarchy compares point measurements with projected point entities from the world model. Differences are used to correct the estimated point entities. Point entities are clustered and connected to form linear entities such as surface boundaries, curves, trajectories, and vertices. Strings of level 1 input events are integrated into level 1 output events spanning a few tens of milliseconds. Level 1 outputs become level 2 inputs.

Level 2 of the sensory processing hierarchy compares observed line entities with projected line entities from the world model. Differences are used to correct the estimated line entities. Line entities are clustered and connected to form surface entities such as planes, terrain features, and surface boundaries. At level 2, data from sensor platform tilt and rotational sensors indicates the direction of gravity and the rotation of the platform. This allows the level 2 world model to transform platform egosphere representations into inertial egosphere coordinates where the world is perceived to be stationary despite rotation of the sensor platform. Acceleration and velocity data from the locomotion system, provide the basis for computing image flow direction. This can be combined with temporal and spatial brightness derivatives to compute depth from image flow. Strings of level 2 input events are integrated into level 3 input events spanning a few hundreds of milliseconds.

Level 3 compares observed surfaces with predicted surfaces. Differences are used to correct the predictions, and groups of contiguous surfaces are clustered into objects. World model knowledge of lighting and texture allow computation of surface orientation, discontinuities, boundaries, and physical properties. Texture and motion of regions provide indication of surface boundaries and depth discontinuities. Strings of level 3 input events are integrated into level 4 input events spanning a few seconds. (This does not necessarily imply that it takes seconds to recognize objects, but that both temporal patterns of motion that occupy a few seconds, and objects, are recognized at level 3. For example, the recognition of the pattern of visual motions produced by a sequence of elementary movements of a manipulator arm might occur at this level.) World model knowledge of the position of the self relative to surfaces enables level 3 to transform from inertial egosphere coordinates into object coordinates.

Level 4 compares observed objects with predicted objects. Correlations and differences between world model predictions and sensory observations of objects allows shape, size, and orientation, as well as location, velocity, rotation, and size-changes of objects to be estimated. Differences are used to correct the world model, and similarities are integrated to characterize groups. World model input from the locomotion and navigation systems allow level 4 to transform object coordinates into world coordinates. Strings of level 4 input events are grouped into level 5 input events spanning a few tens of seconds.

Level 5 compares observed and predicted group characteristics, updates the world model, and integrates similar groups into group2 entities. Strings of level 5 input events are detected as level 5 output events spanning a few minutes. At level 5, the world model map has larger span and lower resolution than level 4.

Level 6 compares observed and predicted group2 characteristics, updates the world model, and integrates clusters of group2 entities into group3 entities, and strings of level 6 input events are grouped into level 6 output events spanning a few tens of minutes. The world model map at level 7 has larger span and lower resolution than at level 6. And so on.

H. Gestalt Effects

When an observed entity is recognized at a particular hierarchical level, its entry into the WM provides predictive support to the level below. The recognition is also passed upward so as to prune the search tree at the level above. For example, a linear feature recognized and entered into the world model at level 2, can be used to generate expected points at level 1. It can also be used to narrow the search at level 3 to entities known to contain that particular type of linear feature. Similarly, surface features recognized at level 3 can generate specific expected linear features at level 2, and limit the search at level 4 to objects with such surfaces, etc. The recognition of an entity at any level thus provides to both lower and higher levels information that is useful in selecting processing algorithms and setting spatial and temporal integration windows to integrate lower level features into higher level chunks.

If the correlation function at any level falls below threshold, the current world model entity or event at that level will be rejected, and others tried. When an entity or event is rejected, the rejection also propagates both upward and downward, broadening the search space at both higher and lower levels.

At each level, the SP and WM modules are embedded in a feedback loop that has the properties of a relaxation process, servo, or phase-lock loop. WM predictions are compared with SP observations, and the correlations and differences are fed back to modify subsequent WM predictions. WM predictions can thus be "servoed" into correspondence with the SP observations. Such looping interactions will either converge to a tight correspondence between predictions and observations, or will diverge to produce a definitive set of irreconcilable differences.

Perception is complete only when the correlation functions at all levels exceed threshold simultaneously. It is the nature of closed loop processes for lock-on to occur with a positive "snap." This is especially pronounced in systems with many coupled loops that lock on in quick succession. The result is a gestalt "aha" effect that is characteristic of many human perceptions.

I. Flywheeling, Hysteresis, and Illusion

Once recognition occurs, the looping process between SP and WM acts as a tracking filter. This enables the WM model predictions to track real world entities through noise, data dropouts, and occlusions.

In the system described here, recognition will occur when the first hypothesized entity exceeds threshold. Once recognition occurs, the search process is terminated or deemphasized, and the thresholds for all competing recognition hypotheses are effectively raised. This creates a hysteresis effect that tends to keep the WM predictions locked onto sensory input during the tracking mode. It may also produce undesirable side effects, such as a tendency to perceive only what is expected, and a tendency to ignore what does not fit preconceived models of the world.

In cases where sensory data is ambiguous, there is more than one model that can match a particular observed object. The first model that matches will be recognized, and other models will be suppressed. This explains the effects produced by ambiguous figures such as the Necker cube.

Once an entity has been recognized, the world model projects its predicted appearance so that it can be compared with the sensory input. If this predicted information is added to sensory input (or multiplied by a positive bias), perception at higher levels will be based on a mix of sensory observations and world model predictions. By this mechanism, the world model may fill in sensory data that is missing, and provide information that may be left out of the sensory data. For example, it is well known that the human hearing system routinely "flywheels" through interruptions in speech data, and fills-in over noise bursts.

This merging of world model predictions with sensory observations may account for optical illusions such as subjective contours and the Ponzo illusion.

V. VALUE JUDGMENTS

Value judgments provide the criteria for making intelligent choices. Value judgments evaluate the costs, risks, and benefits of plans and actions, and the desirability, attractiveness, and uncertainty of objects and events. Value judgment modules produce evaluations that can be represented as value state-variables. These can be assigned to the attribute lists in entity frames of objects, persons, events, situations, and regions of space. They can also be assigned to the attribute lists of plans and actions in task frames. Value state-variables can label entities, tasks, and plans as good or bad, costly or inexpensive, as important or trivial, as attractive or repulsive, as reliable or uncertain. Value state-variables can also be used by the behavior generation modules both for planning and executing actions. They provide the criteria for decisions about which coarse of action to take [45].

Priorities are value state-variables that provide estimates of importance. Priorities can be assigned to task frames so that BG planners and executors can decide what to do first, how much effort to spend, how much risk is prudent, and how much cost is acceptable, for each task.

In animal brains, value judgment functions are computed by the limbic system. Value state-variables produced by the limbic system include emotions. In animals and humans, electrical or chemical stimulation of specific limbic regions (i.e. value judgment modules) has been shown to produce pleasure and pain, as well as more complex emotional feelings such as fear, anger, joy, contentment, and despair [46].

It has long been recognized by psychologists that emotions play a central role in behavior. Fear leads to flight, hate to rage and attack. Joy produces smiles and dancing or leaping-about. Despair produces withdrawal and despondent demeanor. All creatures tend to repeat what makes them feel good, and avoid what they dislike. They attempt to prolong, intensify, or repeat those activities that give pleasure or make the self feel confident, joyful, or happy. They try to terminate, diminish, or avoid those activities that cause pain, or arouse fear, or revulsion.

It is common experience that emotions provide an evaluation of the state of the world as perceived by the sensory system. Emotions tell us what is good or bad, what is attractive or repulsive, what is beautiful or ugly, what is loved or hated, what provokes laughter or anger, what smells sweet or rotten, what feels pleasurable, and what hurts.

It is also widely known that emotions affect memory. Emotionally traumatic experiences are remembered in vivid detail for years, while emotionally non-stimulating everyday sights and sounds are forgotten within minutes after they are experienced.

Emotions are popularly believed to be something apart from intelligence -- irrational, beyond reason or mathematical analysis. The theory presented here maintains the opposite. In this model, emotion is a critical component of biological intelligence, necessary for evaluating sensory input, selecting goals, directing behavior, and controlling learning.

This is not to suggest that intelligent machines can, or should, be endowed with emotions. However, they must have the capacity to make value judgments (i.e. to evaluate costs, risks, and benefits, to decide which coarse of action, and what expected results, are good, and which are bad). Without this capacity, machines can never be intelligent or autonomous. What is the basis for deciding to do one thing and not another, even to turn right rather than left, if there is no mechanism for making value judgments? Without value judgments to support decision making, nothing can be intelligent, be it biological or artificial.

For example, even simple machines typically have sensors that prevent them from injuring themselves by exceeding joint limits, exceeding temperature or pressure limits, or overstressing components. This is analogous to the function of pain sensors in biological creatures.

Intelligent machines may have cost/benefit algorithms that can be used to evaluate plans or steer task execution so as to minimize cost and maximize benefits. There may also be algorithms that evaluate risk and compute the level of uncertainty of knowledge in the world model. Limitations on the amount of memory will require that events and situations evaluated as "important" be remembered, and those evaluated as "unimportant" be cleared from memory. Intelligent machines that include capabilities for learning must have value state-variables that indicate what to "reward" and what to "punish."

Intelligent machines used for military purposes will require means of identifying objects in the world model as "friend" or "foe." If an object is labeled "friend" it should be defended. If it is labeled "foe" it should be attacked. The particular degree of alertness, or tactics chosen for planning tasks in a hostile environment, may be conservative or aggressive based on value state-variables analogous to biological "fear."

A. VJ Modules

Value state-variables are computed by value judgment functions residing in VJ modules. Inputs to VJ modules describe entities, events, situations, and states. VJ value judgment functions compute measures of cost, risk, and benefit. VJ outputs are value state-variables.

The VJ value judgment mechanism can be defined as a mathematical or logical function of the form

$$E = V(S)$$

where

E is an output vector of value state-variables

V is a value judgment function that computes E given S

S is an input state vector defining conditions in the world model, including the self. The components of S are entity attributes describing states of tasks, objects, events, or regions of space. These may be derived either from processed sensory information, or from the world model.

The value judgment function V in the VJ module computes a numerical scalar value (i.e. an evaluation) for each component of E as a function of the input state vector S. E is a time dependent vector. The components of E may be assigned to attributes in the world model frame of various entities, events, or states.

If time dependency is included, the function $E(t+dt) = V(S(t))$ may be computed by a set of equations of the form

$$e(j,t+dt) = (k \, d/dt + 1) \sum_i s(i,t) \, w(i,j)$$

where

e(j,t) is the value of the j-th value state-variable in the vector E at time t

s(i,t) is the value of the i-th input variable at time t

w(i,j) is a coefficient, or weight, that defines the contribution of s(i) to e(j).

Each individual may have a different set of "values," i.e. a different weight matrix in its value judgment function **V**.

The factor (k d/dt + 1) indicates that a value judgment is typically dependent on the temporal derivative of its input variables as well as on their steady-state values. If k > 1 , then the rate of change of the input factors becomes more important than their absolute values. For k > 0, the reduction in negative inputs produces positive output. The more rapid the reduction, the more intense, but short-lived, the reward.

B. Value State-Variable Map Overlays

When objects or regions of space are projected on a world map or egosphere, the value state-variables in the frames of those objects or regions can be represented as overlays on the projected regions. When this is done, value state-variables analogous to comfort, fear, love, hate, danger, and safe will appear overlaid on maps of specific objects or regions of space. BG modules can then perform path planning algorithms that steer away from objects or regions overlaid with fear or danger, and steer toward, or remain close to, those overlaid with comfort or safety. Behavior generation may generate attack commands for target objects or persons overlaid with hate. Protect, or care-for, commands may be generated for target objects overlaid with love.

Projection of uncertainty, believability, and importance value state-variables on the egosphere enables BG modules to perform the computations necessary for manipulating sensors and focusing attention.

Confidence, uncertainty, and benefit state-variables may also be used to modify the effect of other value judgments. For example, if a task goal frame has a moderate benefit variable but low confidence variable, behavior may be directed toward the goal, but cautiously. On the other hand, if both benefit and confidence are high, pursuit of the goal may be much more aggressive.

The real-time computation of value state-variables for varying task and world model conditions provides the basis for complex situation dependent behavior.

VI. CONCLUSION

The amount of research related to intelligent systems is huge, and progress is rapid in many individual areas. Unfortunately, this has not translated into commensurate progress toward a general understanding of the nature of intelligence itself, or even to significantly improved abilities to build intelligent machine systems. Intelligent systems research is seriously impeded by the lack of a widely accepted theoretical framework. Even a common definition of terms would represent a major step forward.

The theory of intelligent systems presented here is only an outline; it is far from complete. Many important issues remain uncertain and many aspects of intelligent behavior are unexplained. Yet, this theory does provide a unifying framework for concepts from a wide variety of disciplines. We hope it will provide insight into how intelligent machine systems can be designed and used in a manufacturing environment.

VII. REFERENCES

1.   L.D. Erman, F. Hayes-Roth, V.R. Lesser, and D.R. Reddy, "The Hearsay-II Speech Understanding System: Integrating Knowledge to Resolve Uncertainty," Computer Survey, Vol. 23, pgs. 213-253, June 1980.

2.   J. E. Laird, A. Newell, and P. Rosenbloom, SOAR: An Architecture for

General Intelligence, Artificial Intelligence, 33 p. 1-64, 1987.

3.   Honeywell Inc., "Intelligent Task Automation," Interim Technical Report II-4, Dec. 1987.

4.   J. Lowerie, et al. "Autonomous Land Vehicle," Annual Report, ETL-0413, Martin Marietta Denver Aerospace, July 1986.

5.   D. Smith, and M. Broadwell, "Plan Coordination in Support of Expert Systems Integration," Knowledge-Based Planning Workshop Proceedings, Austin, TX, December 1987.

6.   J.R. Greenwood, G. Stachnick, H.S. Kaye, "A Procedural Reasoning System for Army Maneuver Planning," Knowledge-Based Planning Workshop Proceedings, Austin, TX, December 1987.

7.   A.J. Barbera, J.S. Albus, M.L. Fitzgerald, and L.S. Haynes, "RCS: The NBS Real-time Control System," Robots 8 Conference and Exposition, Detroit, MI, June 1984.

8.   R. Brooks, "A Robust Layered Control System for a Mobile Robot," IEEE Journal of Robotics and Automation, Vol. RA-2, 1, March 1986.

9.   G. Saridis, "Foundations of the Theory of Intelligent Controls," IEEE Workshop on Intelligent Control, 1985.

10.   A. Meystel, "Intelligent Control in Robotics," Journal of Robotic Systems, 1988.

11.   J.A. Simpson, R.J. Hocken, and J.S. Albus, "The Automated Manufacturing Research Facility of the National Bureau of Standards," Journal of Manufacturing Systems, Vol. 1, No. 1, 1983.

12.   J.S. Albus, C. McLean, A.J. Barbera, and M.L. Fitzgerald, "An Architecture for Real-Time Sensory-Interactive Control of Robots in a Manufacturing Environment," 4th IFAC/IFIP Symposium on Information Control Problems in Manufacturing Technology, Gaithersburg, MD, October 1982.

13.   J.S. Albus, "System Description and Design Architecture for Multiple Autonomous Undersea Vehicles," National Institute of Standards and Technology, Technical Report 1251, Gaithersburg, MD, September 1988.

14.   J.S. Albus, H.G. McCain, and R. Lumia, "NASA/NBS Standard Reference Model for Telerobot Control System Architecture (NASREM)," National Institute of Standards and Technology, Technical Report 1235, Gaithersburg, MD, 1989.

15.   B. Hayes-Roth, "A Blackboard Architecture for Control," Artificial Intelligence, pgs. 252-321, 1985.

16.   J.S. Albus, Brains, Behavior, and Robotics, BYTE/McGraw-Hill, Peterbourough, N.H.,1981.

17.   G.A. Miller, "The Magical Number Seven, Plus or Minus T: Some Limits on Our Capacity for Processing Information," The Psychological Review, 63, pp.71-97, 1956.

18.   A. Meystel, "Theoretical Foundations of Planning and Navigation for Autonomous Robots," International Journal of Intelligent Systems, 2, 73-128, 1987.

19.   E.D. Sacerdoti, A Structure for Plans and Behavior, Elsevier, New York, 1977.

20.   M. Minsky, "A Framework for Representing Knowledge," in Winston (1975), 211-277.

21.   R.C. Schank and R.P. Abelson, Scripts Plans Goals and Understanding, Hillsdale, NJ, Lawrence Erlbaum Associates, 1977.

22.   D.M. Lyons, and M.A. Arbib, "Formal Model of Distributed Computation Sensory Based Robot Control," IEEE Journal of Robotics and Automation in Review, 1988.

23.   D. W. Payton, "Internalized Plans: A Representation for Action Resources," Robotics and Autonomous Systems, 6, 89-103, 1990.

24.   A. Sathi, and M. Fox, "Constraint-Directed Negotiation of Resource Reallocations," CMU-RI-TR-89-12, Carnegie Mellon Robotics Institute Technical Report, March, 1989.

25.   J.J. Gibson, The Ecological Approach to Visual Perception, Cornell University Press, Ithaca, N.Y. 1966

26.   H. Samet, "The Quadtree and Related Hierarchical Data Structures," Computer Surveys, 16-2, 1984.

27.   P. Kinerva, Sparse Distributed Memory, MIT Press, Cambridge 1988.

28.   J.S. Albus, "Data Storage in the Cerebellar Model Articulation Controller (CMAC)," Transactions ASME, September 1975.

29.   M. Bradey, "Computational approaches to image understanding," ACM Computing Surveys 14, March, 1982.

30.   T. Binford, "Inferring surfaces from images," Artif. Intell 17, 205-244, 1981.

31.   D. Marr and H.K. Nishihara "Representation and recognition of the spatial organization of three-dimensional shapes," Proc, R. Soc, Lond. B 200, 269-294, 1978.

32.   R.F. Riesenfeld, "Applications of B-spline Approximation to Geometric Problems of Computer Aided Design," Ph.D Thesis, Syracuse University (1973). Available at University of Utah, UTEC -CSc-73-126.

33.   J.J. Koenderink, "The Structure of Images," Biological Cybernetics, 50, 1984.

34.   D. Marr, *Vision*, W.H. Freeman, San Francisco, 1982.

35.   D.H. Hubel and T.N. Wiesel, "Ferrier Lecture: Functional architecture of macaque monkey visual cortex," *Proc. Roy. Soc. Lond. B. 198*, 1-59 (1977).

36.   J.H.R. Maunsel and W.T. Newsome, "Visual processing in monkey extrastriate cortex," Ann. Rev. Neurosci. 10: 363-401, 1987.

37.   J.S. Albus and T.H. Hong, "Motion, Depth, and Image Flow," Proceedings of the IEEE Robotics and Automation, Cincinnati, OH, 1990.

38.   D. Raviv, and J.S. Albus, "The computation of range from image flow given knowledge of eye translation and rotation," IEEE Transactions on Robotics (to be published).

39.   E.W. Kent and J.S. Albus, "Servoed World Models as Interfaces between Robot Control Systems and Sensory Data," Robotica, Vol. 2, pags. 17-25, 1984.

40.   E.D. Dickmanns, T.H. Christians, "Relative 3D-State Estimation for Autonomous Visual Guidance of Road Vehicles," Intelligent Autonomous System 2(IAS-2), Amsterdam, 11-14 December, 1989.

41.   A.L. Yarbus, *Eye Movements and Vision*, Plenum Press, 1967.

42.   R. Bajcsy, "Passive perception vs. active perception," Proc, IEEE Workshop on Computer Vision, Ann Arbor, 1986.

43.   Y.L. Grand, *Form and Space Vision*, Table 21, Indiana University Press, Bloomington, 1967

44.   J.S. Albus, "Outline for a Theory of Intelligence," IEEE Transactions SMC, (to be published).

45.   G.E. Pugh, *The Biological Origin of Human Values*, Basic Books, New York, 1977.

46.   A.C. Guyton, *Organ Physiology, Structure and Function of the Nervous System*, 2nd ed., Philadelphia: W.B. Saunders, 1976.

SELECTED TOPICS IN INTEGRATION
FROM THE AMRF

ALBERT JONES

Deputy Program Director, AMRF
Manufacturing Engineering Laboratory
National Institute of Standards and Technology
Gaithersburg, Md 20899

I. INTRODUCTION

When American industry began to move toward automated manufacturing systems, the National Institute of Standards and Technology (NIST) (then the National Bureau of Standards) had a responsibility to help. To better understand how NIST could help, its researchers asked three questions:

1) What was needed to make measurements in such an automated factory?

2) What was needed to make it possible for companies to move toward automation in a modular and incremental fashion? and,

3) What was needed to handle all of the data and information that would be generated and used by the computers in these automated systems?

Our researchers decided that the answers to each of these questions (whatever they turned out to be) would fall into two major categories: new standards and new technologies.

In the late 70s, two projects were initiated, which began a long term commitment to finding answers to these questions. The first was on automated measurement techniques. This was described in the preceding chapter. The second was on a control system for robots. This is described in a later chapter. During the evolution of these projects, we soon recognized that this was not enough. In order to really understand the standards and technology problems that would be faced, we were going to have to build a small automated manufacturing system.

To meet this challenge, the Center for Manufacturing Engineering at NBS conceived of the Automated Manufacturing Research Facility (AMRF) [1]. The AMRF would be designed to provide a forum for conducting research into standards- and technology- related problems in metrology and integration. Industry, academia, and other government agencies would play a significant role in the development of the AMRF through direct appropriations, equipment loans, and cooperative research programs. The focus of the AMRF was to be small-batch, discrete-parts, metal fabrication.

Physically, the AMRF was to contain several robots, machining centers, inspection equipment, storage and retrieval systems, wire-guided vehicles, and numerous computers. The exact nature and configuration of these components has changed dramatically since the first AMRF public demonstration in 1983. Figure 1 shows the 1983 configuration and Figure 2 shows the 1987 configuration. Changes are continuing to this day.

Realizing that they would have to deal with this diversity of components and changing shop-floor configuration, AMRF researchers have been in a unique position to address questions 2 and 3. The AMRF view is that the key to answering these two questions lies in the design and implementation of a global architecture. Once such an architecture is in place, even if it changes over time, both standards and technology requirements become clearer and modularization and integration become possible.

From our experience in building the AMRF, we are convinced that the process of designing and implementing such an architecture is an iterative one. We are also convinced that such a global architecture must integrate separate architectures for production, information, and communications management. This was and continues to be a dramatically different, and from our experience superior, approach to the monolithic one advocated by many others.

In this chapter we describe these architectures. We provide details on the principles employed in generating their designs and discuss some experiences gained during their implementation. Much of its contents has been excerpted from papers published previously by a variety of NIST researchers. These papers are referenced in each section. The period covered extends from approximately 1980 through 1988.

II. PRODUCTION MANAGEMENT

Production management functions, as we use the term, can be divided into three groups: manufacturing data preparation, production planning/inventory control,

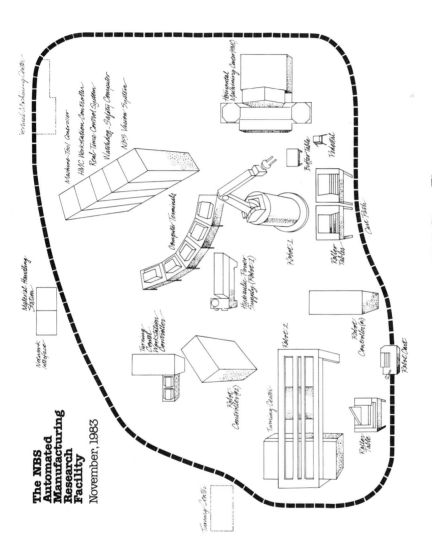

**The NBS
Automated
Manufacturing
Research
Facility**
November, 1983

Vertical Machining Center

Machine-Tool Controller
HMC Workstation Controller
Real-Time Control System
Watchdog Safety Computer
NBS Vision System

Horizontal Machining Center (HMC)

Pedestal

Buffer Table

Computer Terminals

Robot 1

Roller Tables

Cart Path

Hydraulic Power Supply (Robot 1)

Material Handling Station

Network Interface

Turning Center Workstation Controller

Robot Controller (#2)

Robot 2

Robot Controller (#1)

Robot Cart

Turning Center

Roller Table

Turning Center

Figure 1. 1983 AMRF Shop Floor

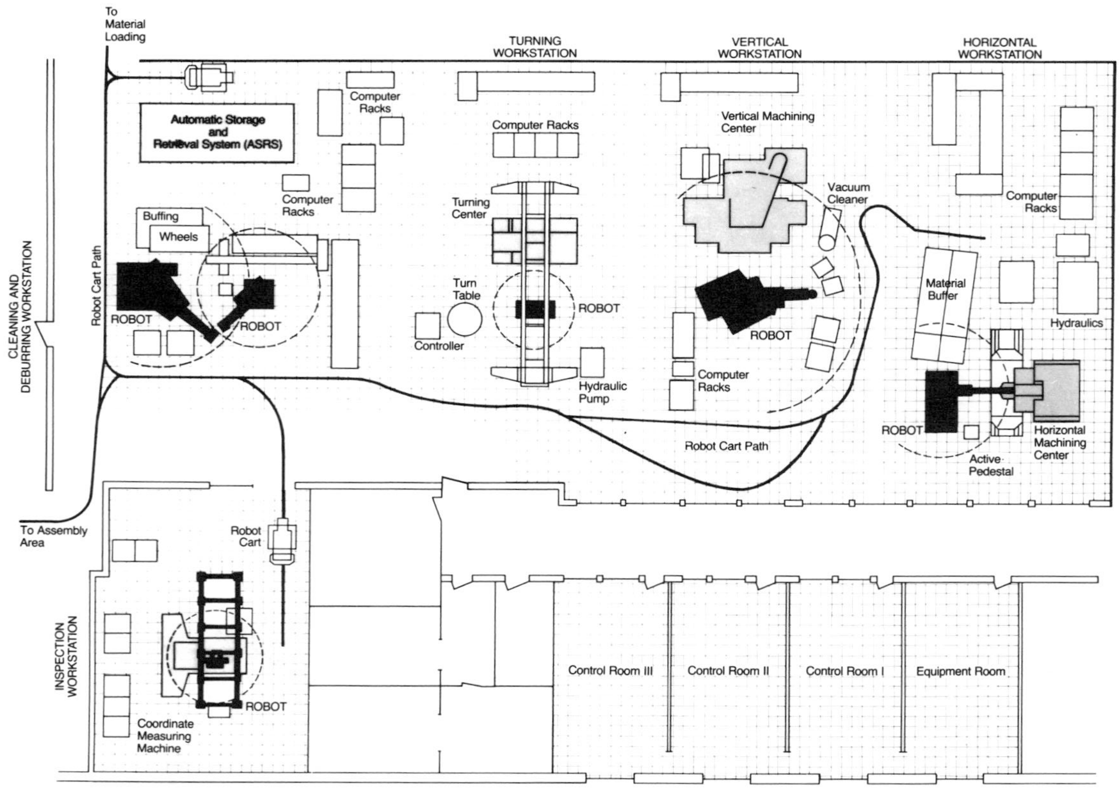

Figure 2. 1987 AMRF Shop Floor

and shop floor control. To date, the AMRF has concentrated on the first and last of these.

A. Manufacturing Data Preparation

In 1987, the AMRF began a project called Manufacturing Data Preparation (MDP). MDP brought together several existing AMRF projects in an attempt to integrate those functions which generate and update the data needed to drive shop floor control functions. Included are design engineering, process planning, NC programming, robot programming, and inspection programming. Table 1 shows the way these functions are carried out today and the way, we believe, they will be carried out in the future.

Table I. Evolution of MDP functions

MDP TODAY	MDP TOMORROW
Human Beings	Computers
Computer Assisted	Human Assisted
Off-Line	On-Line
No Interaction	Integrated System

Currently, as indicated in the table, MDP consists of many human-intensive, computer-assisted, activities. These activities are typically carried out far in advance of when their output is actually needed. Moreover, there is little or no exchange of ideas or information (before or after the fact) among the design engineers, process planners, and manufacturing engineers who perform those functions. The trend toward "Just-in-Time" manufacturing and "Concurrent Engineering" is providing the impetus to change all of this. There is a big push to automate and integrate MDP functions as much as possible with computers doing most of the work and human beings reviewing the decisions. In addition, there is an attempt to begin the execution of these functions closer to the time that part production is expected to begin. This gives the decision-makers more accurate information about the state of the manufacturing equipment that they may wish to use. In addition, it is expected to decrease dramatically the probability of selecting a piece of equipment which is either over-capacitated or down for repair. This can reduce shop floor congestion and improves plant performance.

The input to MDP consists of a complete part description, a portion of which is used by all of the functions. An international effort has been underway for several years to develop a standard product description. In the United States, this effort is called Product Data Exchange using STEP (PDES). STEP is the international standard under development and stands for the STandard for the

Exchange of Product model data [2]. An initial attempt to provide such a description was undertaken in the AMRF. It was called the AMRF Part Model. These efforts are described more fully in a later chapter in this book.

All of the internal MDP functions will be performed by commercial products. Some are fully automated, but most are not. This creates a major integration problem - data exchange - which we have focused on for a number of years.

The output from MDP that is used most frequently by controllers in the shop floor system is the process plan (see next section). It contains the information needed to manufacture, transport, and inspect parts. In the long run, all AMRF shop floor controllers will use process plans to determine how to plan, schedule and execute assigned jobs. This requires several changes to existing process plans. First, they must have a multi-level structure which parallels the control hierarchy (see next section). Second, process plans at each level must provide alternate processing sequences, with precedence relations. Third, plans at different levels must have the same general structure (AND/OR graphs are one possibility). This simplifies the software development and allows for standardization of both content and "format". Finally, since computers will be responsible for processing it, this information must be provided in a consistent, error-free, and machine-readable form. AMRF researchers have developed such a representation, called ALPS [3].

B. Shop Floor Control

The AMRF shop floor control architecture is based on a multi-level, hierarchical structure. Each module in this structure has one supervisor, but may have many subordinates. The design of this hierarchy is based on three guidelines which grew out of earlier work on robot control systems [4]:

1) Levels are introduced to reduce complexity and limit responsibility and authority,

2) Each level has a distinct planning horizon which decreases as you go down the hierarchy

3) Decision-making and control reside at the lowest possible level.

In designing an architecture based on these principles, the AMRF shop floor control hierarchy has gone through four major stages.

1. Stage 1 - The Original Architecture

The first architecture contained five levels: facility, shop, cell, workstation, and equipment (see Figure 3). Only the bottom three levels were implemented. A brief discussion of the responsibilities assigned to each of those levels follows. More details can be found in [1].

The facility level was responsible for implementing the front office functions

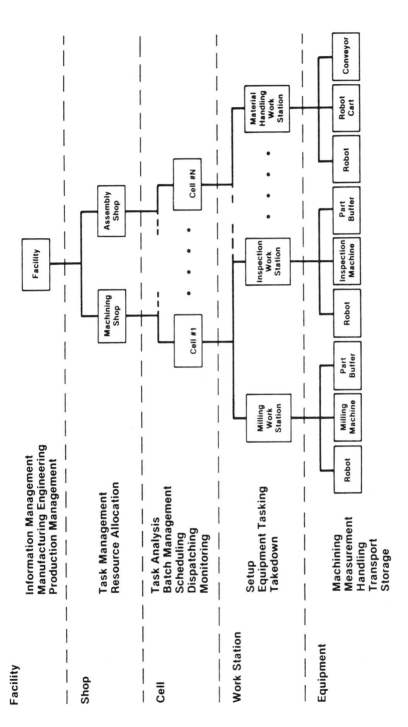

Figure 3. Original AMRF Control Hierarchy

that are typically found in manufacturing facilities. Activities at this level were grouped into subsystems that fell into three major functional areas: manufacturing engineering, information management, and production management. Manufacturing engineering functions included the MDP functions described above. The information management activities were to provide the user-data interfaces to support administrative and business management functions. Production management was to track major projects, generate long-range schedules, identify production resource requirements, determine the need for additional capital investments to meet production goals, determine excess production capacity, and summarize quality performance data.

The shop level was responsible for planning and coordinating the production and support jobs on the shop floor. It was also responsible for allocating resources to those jobs. Two major component modules were identified within the shop controller: a task manager and a resource manager. The task manager was responsible for capacity planning, grouping orders into batches, assigning and releasing batch jobs to cells, and tracking individual orders to completion. The resource manager had responsibility for allocating the production resources to individual cells, managing the repair of existing resources, and ordering new resources.

At the cell level, batch jobs of similar parts were sequenced through workstations, and supervision provided for various other support services, such as material handling and inspection. The cell was designed to bring some of the efficiency of a flow shop to small batch production by using a set of machine tools and shared job setups to produce a family of similar parts. The AMRF cells were originally conceived to be dynamic production control structures which allowed time-sharing of workstations.

The activities of small integrated physical groupings of shop floor equipment were directed and coordinated at the workstation level. A typical AMRF workstation consists of a robot, a machine tool, a material storage buffer and several control computers. Machining workstations process trays of parts that are delivered by the material handling system. The workstation controller sequences equipment-level subsystems through job setup, part fixturing, cutting processes, chip removal, in-process inspection, job takedown, and cleanup operations.

AMRF equipment controllers are front-end systems that are closely tied to commercial equipment or industrial machinery on the shop floor. Equipment controllers are required for robots, numerically-controlled machine tools, coordinate-measuring machines, delivery systems, and storage/retrieval devices. Equipment controllers perform two major functions: (1) they translate workstation commands into a sequence of simple tasks that can be understood by the vendor-supplied controller, and (2) they sequence and monitor the execution of these tasks via the sensors attached to the hardware.

2. Stage 2 - Moving Toward A Generic Controller

After the initial implementation of this architecture in 1983, a group of AMRF
researchers took another look at the functionality of the cell and workstation
controllers. They concluded that there was a common set of activities that each
of the modules performed:

1) The decomposition of complex jobs into a set of simpler tasks
2) The development of networks describing task dependencies
3) The assignment of tasks to subordinates
4) The allocation of resources to tasks

These activities are very similar to those typically performed by a project
manager. AMRF researchers believed that it might be possible to design a
single structure which could be utilized at each level of the hierarchy. The first
design [5] of such a controller had three functional modules: Production
Manager, Queue Manager, and Dispatch Manager. It was designed and
implemented at the cell level.

a. Production Manager: The Production Manager (PM) received a list of jobs
to process and resource allocations from its supervisor and feedback from the
Queue Manager, indicating the current status of all in-process work. Whenever
the PM selected the next job from this list to enter production, it would retrieve
all process plans related to that job and initialize each step in the process plans
as a run-time task. The initialization process included the creation of a task
record in the database for each step in the process plan, the assignment of a
unique identification number to the task, the creation of cross reference pointers
to other precedent or subsequent tasks, and the computation and entry of any
initial timing estimates. Finally, a real-time production plan was generated. This
production plan contains:

1) A list of tasks assigned to each subordinate
2) A set of precedence and coordination flags for each task
3) An exact execution sequence for the tasks
4) All timing data for each task

The feedback information from the Queue Managers (QM) was used to
determine the accuracy of the current plan. If everything is on schedule, then
the plan is not altered. But, whenever a problem is encountered, the current
plan must be changed to include the expected effects of any delays. The nature
of the problem and the current state of the system would determine whether the
supervisor must be informed before any corrective action is taken. In either
case, the Queue Managers must be informed of any changes in the plan.

b. Queue Manager: In the original formulation, there was one Queue Manager
(QM) for each subordinate controller. It received a list of jobs from the

Production Manager and placed them in the job queue in the assigned order. Each such job was assigned an ACTION value. Whenever a job received an EXECUTE, it was passed down to Dispatch Manager (DM) for release to the appropriate subordinate. Whenever the DM sent feedback indicating that a job was completed, the Queue Manager removed this entry from the queue and reported this up to the PM. The QM had no capability to resolve any problems, other than those relating to scheduling or timing, that are reported by the Dispatch Manager. Information on operational problems was simply passed up to the PM where the current schedule and system state would be used to select the appropriate action.

c. Dispatch Manager: There was one Dispatch Manager (DM) for each subordinate controller. The DM had a library of software routines that support the internal processing of the tasks, previously derived from the process plan, that are specified for each task. The routines at this level defined the way in which the DM, as a supervisor, managed the execution of the primitive work element by issuing commands to and receiving feedback from the Production Manager level of that subordinate. No planning functions were identified for the DM. It released jobs at the appropriate time and monitored the execution of those jobs to completion. Any problems that resulted in delaying completion times were reported directly to the Queue Manager. Although it was possible to combine the QM and DM into a single job manager, this aggregation would have increase the complexity of the QM unnecessarily.

3. Stage 3 - Introducing Distributed Decision-Making

After implementing the production control module at the cell level, AMRF researchers realized that similar decision-making was necessary at every level in the hierarchy. An attempt was made to provide much more detail about those decisions. What follows is described more fully in [6].

a. Facility Level: The facility level had sole responsibility for the business and strategic planning functions which support the entire manufacturing enterprise. Better mathematical models were required to aid top management in assessing and justifying the potential benefits and costs of flexible automation. In addition, once the decision had been made to employ advanced technology, new techniques were needed in cost accounting, depreciation, capital investment strategies, and many other business functions. Existing methodologies were, and still are, unable to measure the impact of this flexibility in a meaningful way.

Another responsibility of the facility level was the manufacturing-data-preparation described above. Schedules must be generated for all of the activities required to complete those MDP functions, particularly as they were being moved closer to the actual production. These schedules included both new customer requests and revisions to existing data required by changing conditions on the shop floor. In addition, new methods were needed to aid in the classification and coding of parts from CAD data, geometric modeling, decomposition of complex geometries into primitive features that can be machined and inspected, and the design, revision, and verification of process plans.

b. Shop Level: The shop level received a list of customer requests and any assigned priorities or due dates from the facility level. The shop level sequenced these requests, grouped them into batches, and determined the order in which they would be released to the manufacturing cells on the shop floor. It then produced a schedule which included the cells to be used for each batch, estimated start and finish times at each cell, and the required material transfers among those cells. These plans would have to be updated any time a new request was issued, an existing request was cancelled or given a higher priority, or a significant problem occurred.

The shop also had overall responsibility for inventory control, tool management, capacity planning, and preventive maintenance for all equipment in the shop. These activities must be managed to support the schedules developed at this level. An important issue to be resolved at the shop level was the future use of existing techniques such as material resource planning and master production scheduling. In an environment like the AMRF, in which decisions are pushed down to the lowest level, these global planning approaches were no longer applicable.

c. Cell Level: A cell controller must coordinate the activities of its subordinate workstations to complete the jobs assigned by the shop. Each job required the services of one or more workstations including material handling and had a due date and priority associated with it. The cell would sequence these jobs and develop a schedule of anticipated start and finish times and priorities for each job at each workstation. It determined which workstations were needed and the order in which they would be visited. It would also arrange for the requisite material transfers in support of that schedule. When conflicts or delays were reported by a workstation controller, the cell would be expected to replan, reroute, and reschedule to overcome them.

Coordinating the activities at these workstations becomes more difficult when there exist shop-wide, shared resources like material transport devices.

d. Workstation Level: Workstation controllers were responsible for coordinating the activities of subordinates to execute the tasks assigned by a cell controller. Although the exact nature of the tasks were workstation-dependent, they typically consisted of receiving materials, shipping materials, setup, takedown, machining/inspecting a list of features. The workstation controller generated a sequence in which to perform these tasks and a schedule for each of its subordinates.

In addition to the aforementioned problems, the material-handling workstation controller had several other problems that it had to address. These special problems are directly related to its primary responsibility of planning and coordinating the activities required to move trays of materials around the factory. It must locate the material, assign a transportation device (or devices) to pick up and deliver that material, and determine the routes it will follow in executing the task. Further, all these activities must be coordinated and monitored for possible changes and updates.

Assigning trays to batches of parts must also be addressed. This problem is complicated in an environment in which a batch size of one or two is the rule rather than the exception. In this case, a single tray could contain several batches of parts, each having a different geometry. Further complications are that deliveries to more than one workstation may be combined on a single tray and that each transportation device may be capable of carrying more than one tray.

e. Equipment Level: The lowest level in the hierarchy contains the AMRF-built equipment controllers. There were three classes of equipment in the AMRF: stationary robots, machining centers, and material storage/retrieval/ transport devices. The mathematical decision problems to be solved by each equipment controller fall into two major categories. The first is sequencing. Each controller must sequence through the current tasks assigned by its supervisory workstation. They may be rank-ordered, with expected completion times associated with each task. The second set of problems is equipment-dependent, and is discussed in more detail in [6].

4. Stage 4 - Integrating decision-Making and Control

Recently, some AMRF researchers have proposed a more general architecture which integrates the control ideas described in II.B.1 with the decision-making requirements described in II.B.2 and II.B.3. The decisions described above can be divided into two classes: planning the activities to be carried out and scheduling their execution. That is, given the information provided by MDP, they determine exactly how and when to execute their assigned tasks. Control then attempts to implement those decisions. This involves monitoring subordinates and recovering from minor problems.

The number of levels in this architecture is no longer fixed at five. It is based on the decomposition of the corresponding mathematical formulations of these decision-making and control problems. This mathematical decomposition results in a decomposition of decision space, not physical space. It is this new type of "spatial decomposition" that determines the number of levels in the shop floor hierarchy not the physical arrangement of the equipment on the shop floor.

Each module in this new hierarchy is an "expanded" production control module performing four major functions (see figure 4): assessment, optimization, execution, and monitoring. These are described briefly below. More details can be found in [7].

a. Assessment Function: The assessment function generates a run-time production plan which will enable each job to be completed within the specified time limits. It is also responsible for changing an existing plan when conditions

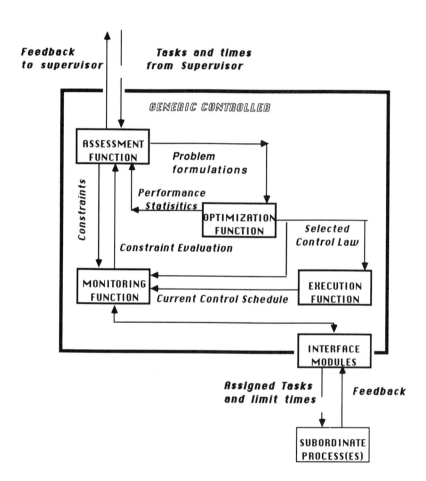

Figure 4. A Generic Control Module

make it no longer feasible (see Figure 5). This run-time production plan will be one of the alternatives contained in the process plan generated by MDP. It will contain 1) the list of tasks assigned to each subordinate, 2) precedence relations among those tasks, and 3) proposed durations for each task.

For a new job, this involves several steps. First, the assessment function retrieves the process plan previously generated by MDP. The process plan contains all the possible alternatives for making the part. It then will select feasible ones, based on the current state of the system. This set of feasible alternatives is passed to the Optimization function where they are ranked by one or more performance criteria. These rankings are used to select the run-time plan. It then requests the optimizer to schedule the part using the selected run-time plan. It will calculate the start and finish times for the job based on that plan and pass them up as feedback to the supervisor. The run-time plan is put into the database for later use by the assessment function of subordinates.

The feedback information from the Optimization function is used to determine the viability of the current production plan. Whenever shop floor conditions evolve to the point that supervisory constraints cannot be met with the current production plan, a new one must be created. The assessment function must now specify a new set of a candidate production plans to meet these constraints. This, of course, may not be possible. When this happens, it will be necessary to negotiate a new set of limit times with the supervisor.

b. Optimization Function: The optimization function (the optimizer) performs three major activities (see Figure 6). It evaluates proposed production plans from the assessment function. It generates a list of tasks and corresponding start/finish times for subordinates - a schedule. Finally, it resolves, if possible, any conflicts and problems with the current schedule identified by the monitoring function.

As discussed above, the optimizer evaluates alternative production plans for each job. This analysis is performed to determine the impact that each plan would have on the rest of the workload (i.e., the active schedule if it were selected. This impact can be analyzed in terms of one or more performance measures passed down from the assessment function. They can include tardiness of current jobs, utilization of subordinates, load on the system, and throughput, among others. The optimizer will rank these alternatives and pass the results back to the assessment function which makes the final determination.

Once this selection has been made two things happen. First, the optimizer schedules (computes estimates for start and finish times) all of the tasks in that plan. The performance measures used in the scheduling analysis must be consistent with those used in generating the plan. A scheduling rule is found which optimizes those performance measures. The resulting schedule is then used by the assessment functions of each subordinate in generating their run-time plans. These times are also passed up to the assessment function so that it can update its own estimates of job completion.

Second, this plan provides the tasks which are assigned to the subordinates. The dynamic evolution of these subordinate systems can cause delays which

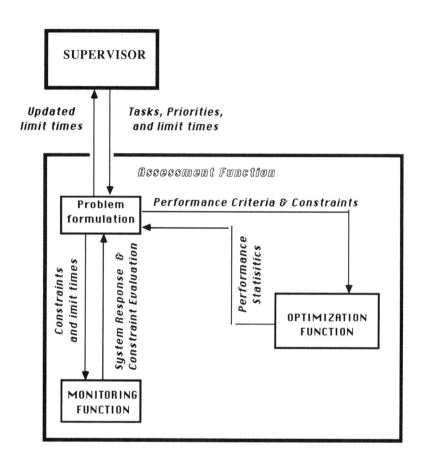

Figure 5. The Assessment Function

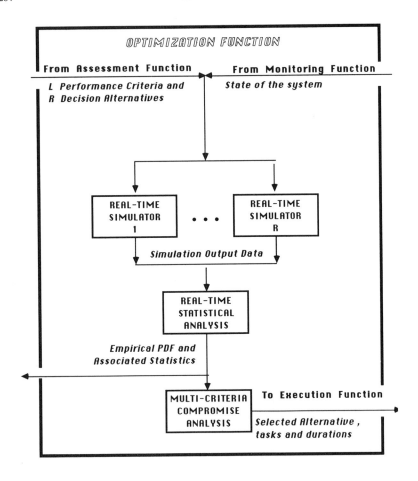

Figure 6. The Optimization Function

make the schedule generated by the optimizer infeasible. The monitoring function detects such a situation and invokes the optimizer to resolve any conflict as quickly as possible. A two step process is envisioned. First, the impact of the delay must be determined The outcome of this analysis determines the viability of the current objectives, scheduling rule, and run-time plan. They remain viable as long as there is enough slack in the original schedule to absorb the ripple effect of the delay. The optimizer can change objectives and scheduling rules, but it cannot change the current run-time plan. In either case, a new schedule can easily be generated using the method described in [7]. Whenever rescheduling is insufficient, a new run-time plan is required. The procedure described above is followed. A new plan is selected, which eventually results in a new schedule.

c. Execution Function: The Execution function controls the execution of both the plan and schedule generated by the Optimization function. In particular, it 1) computes the expected start and finish times for each subordinate task in the selected plan using the selected scheduling rule, and 2) tries to restore feasibility using that plan and rule whenever violations of those times arise from the decisions/actions of subordinates (see Figure 7). The former is done using a single-pass simulation initialized to the current state of the system. The error recovery strategy under development is explained in detail in [7].

Once the scheduled start and finish times are computed, they are passed (along with the tasks) to the subordinates where refined schedules are computed, and to the Monitoring function to be used for identifying infeasibilities.

d. Monitoring Function: The monitoring function has two major responsibilities: updating the system state from the feedback provided by subordinates and identifying constraint violations based on the current state (see Figure 8).

The system state summarizes both the operational status of subordinates and the status of all tasks assigned to those subordinates. This state is an aggregation of all the feedback information provided by subordinates. This state is then used to determine whether or not any of the constraints imposed by the assessment function has been violated. If this is the case, the execution function will use the selected plan and scheduling rule to try and restore feasibility. If it cannot, the optimization function will try to restore feasibility by selecting another scheduling rule/production plan from the feasible set provided by the assessment function. If it cannot, then the assessment function will either update the feasible plans and rules or negotiate with the supervisor to relax some constraints.

C. Future Evolution

One of the most difficult problems faced during integration testing of the AMRF was equipment failure. To address this, a new project called Manufacturing Systems Integration began in 1989. The goal is to develop implementation plans

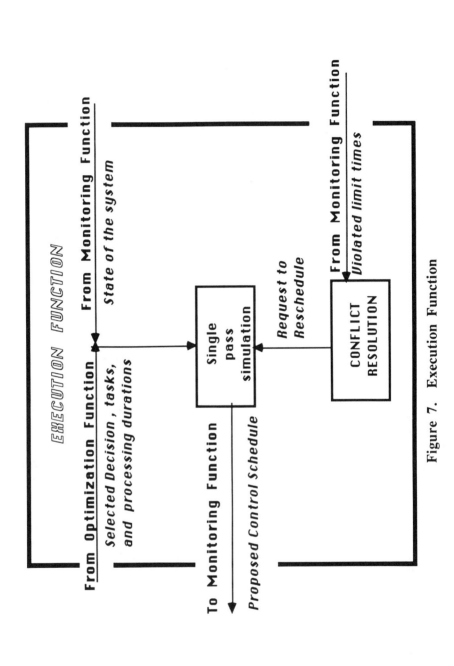

Figure 7. Execution Function

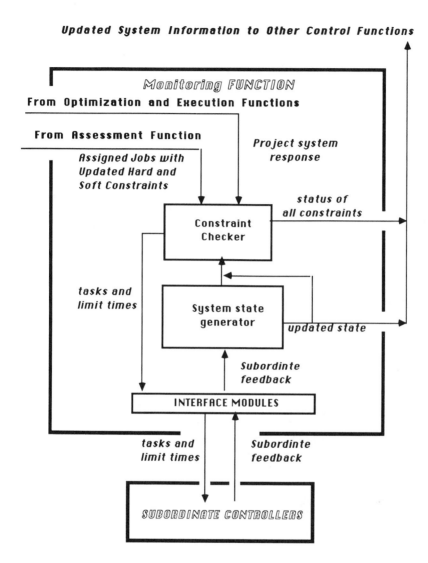

Figure 8. Monitoring Function

for a new "table-top" AMRF. This table-top AMRF will 1) be based on the architecture presented in the II.B.4, and 2) have no physical equipment to control. All equipment will be emulated, simulated, or animated. This will provide the greatest flexibility in testing a variety of architectural concepts and the integration of commercial software packages.

III. ISSUES IN DATA MANAGEMENT

A. Overview

The purpose of a data management system in an AMRF-type manufacturing environment is provide users (application programs in the production management architecture) with timely access to all essential data. Our experience has led us to the conclusion that it is essential to design and implement a data management architecture which is separate from, but integrated with, the production management architecture. This allows the two structures to be developed independently, provided their interrelationships are well understood. There are, however, many characteristics of such an environment which make this approach difficult to implement. Barkmeyer et al. [8] have discussed this at length, and we now present a brief discussion of these characteristics.

1. Heterogeneous System Environment

The computers and production equipment which make up these advanced manufacturing systems will be purchased from a variety of vendors over a long period of time. This implies that data is likely to be physically distributed across a network of heterogeneous computers. These local repositories will have a wide range of data access, storage, management, and sharing capabilities. The data management architecture must make these differences transparent to users; i.e., users simply make requests and receive data. In addition, users should not be concerned about the effort required to satisfy their requests. To achieve these goals, the data system must provide users with a common method of accessing information. The data system must deal with the problem of translating requests in this common form to operations on the underlying data repositories, wherever and whatever they may be. This results in transmission and translation of component operations to the appropriate database management systems, assembly of information from multiple sources, and conversion of the information to the form the user expects.

2. Real-Time Operations

A variety of data is used by computer systems that control shop-floor equipment to make real-time decisions. If that data is not present when it is needed, erroneous decisions or no decision may be made, resulting in processing delays and reduced plant throughput. To complicate matters, some of that data may be shared by several users with different "real-time" access requirements. This implies that the data system must enable asynchronous interchanges of information between production processes. This in turn requires the replication of some information units on two or more systems and the frequent and timely updates of those units.

3. Data Delivery and Job Scheduling

Highly-automated systems are highly dependent on electronic information. It is important to realize that data delivery, like material delivery, takes time and must be included in the planning of each job. Actual part production cannot start until all of the required information is transferred to the computer responsible for controlling the process performing that production. The notion that this transfer is effectively instantaneous is becoming obsolete as the speeds of automated systems themselves increase. This means that data is quickly becoming a critical resource which must be scheduled. Poor "data scheduling" will lead to the same delays, bottlenecks, and idle equipment that arises from poor "job scheduling." This implies the need for coordination between the data scheduling function in the data management architecture and its counterpart in the production management architecture.

B. What Constitutes an Architecture?

To meet the requirements and constraints described above, a data management architecture must address three major concerns: data modeling, database design, and data administration.

1. Data Modeling

Developing a conceptual model of all the information involved in the entire production management spectrum is critical to the success of any integrated data management system. Because the amount of information is so large, a divide-and-conquer approach to performing the analysis must be taken. Experts on individual functions in the production management architecture will perform the analysis and develop a conceptual information model for each functional area. This results in models for product data, process plans, work-in-process, resource capabilities, etc. Then the resulting component models must be integrated into a single enterprise model. The enterprise model is the conceptual representation of the global information base.

Several powerful modelling techniques now available allow representation of the real-world objects themselves, as well as the information units which describe and distinguish them. Using such a technique, one can distinguish a concept from its computer representation. This avoids problems with multiple representations of the same concept or similar representations of somewhat different concepts. This capability is vital to the development of an integrated data model.

From this enterprise model, it is possible to extract subsets which represent "views" of the global information base possessed by individual production management processes. But, the processes actually want to use data, and they want that data organized in a specific way, which we call an external view. An external view is an interchange representation and is one of the elements of the interface between data management functions and the production management functions. Generating these views correctly is a matter of considerable current research interest. They are extremely important because they make the abstract objects disappear and the modelled information units acquire specific physical representations.

2. Database Design

After the global enterprise model is completed, we must then map this model onto live databases. This process is called database design. It must result in databases which are consistent with the model and tuned to the timing and access requirements of the production management functions that use them. Because of the evolutionary nature of manufacturing systems, it is not possible to start this process from scratch. There are vendor-supplied databases and data systems, which are difficult to alter or augment. There are also "legacies" - large reservoirs of previously developed information which already have an imposed organization. The only solution which meets production management requirements and the legacy and autonomy constraints is to divide up the data itself into multiple databases serving specific production functions well. Since much of that data must be shared, two problems result:

1) partitioning - some production functions must simultaneously access information stored in two or more databases, and /or

2) replication - some data must be stored simultaneously in two or more different databases, and maintained consistently.

The available options for the placement of databases in the computer system complex, and for the selection of specific data management systems to support them, are dictated to a large extent by the architecture of the "global" data administration system.

3. Data Administration

The administration portion of the data management architecture provides the data services controlling access to all data:

1) query processing, which is concerned with the command/response interface to user programs, the data manipulation language and its

interpretation, and the validation of user transaction requests,

2) transaction management, which is concerned with identification of databases participating in a given transaction, transaction scheduling and conflict resolution, and

3) data manipulation, execution of the operations on the databases.

There are three control architectures for data administration systems which have been used with varying degrees of success in various business applications: centralized data and control, distributed data and control, and hybrid systems.

The totally centralized approach is the traditional design, the simplest, and the most workable. Whether this is a feasible architecture for manufacturing in the long-run is unresolved. There are currently available high-speed, internally redundant, fault-tolerant, integrated centralized systems. But even if such systems can keep pace with the growing demands and time-constraints of automated production systems, the centralized architecture is not workable from the point-of-view of subsystem autonomy. Vendors of design and planning systems, for example, cannot assume that every customer will have such a facility, and will therefore develop local databases and data services to meet the needs of the products they provide. Consequently, the nominally centralized architecture will in fact consist of a collection of autonomous systems copying information to and from a single centralized facility, according to some externally-specified plan. At best, if the external plan provides uniformly for concurrency control, security and the like, what results is centralized data with distributed control.

The canonical architecture for the totally distributed approach (Figure 9) consists of local data management systems which process locally originated and locally satisfiable requests and negotiate with each other to process all other requests. In this case, difficult problems of concurrency control, distributed transaction sequencing and deadlock avoidance occur and must be resolved by committee. While there has been considerable research in these areas, satisfactory solutions have not been found. Moreover, the dynamic evolution of manufacturing systems leads to the problem of configuration changes in the complex, which requires informing all existing participants and modifying their distribution information.

The hybrid architecture attempts to combine the best features of both centralized and distributed architectures. Subsystem autonomy and high throughput are achieved by allowing local data systems to process locally originated operations on local data (see Figure 10). Operations which transcend the scope of a local system are sent over the network to a centralized "global query processor" for distribution to the appropriate sites. The global query processor acts as a central arbiter for resolving the characteristic problems of distributed transactions and for handling configuration changes in the data

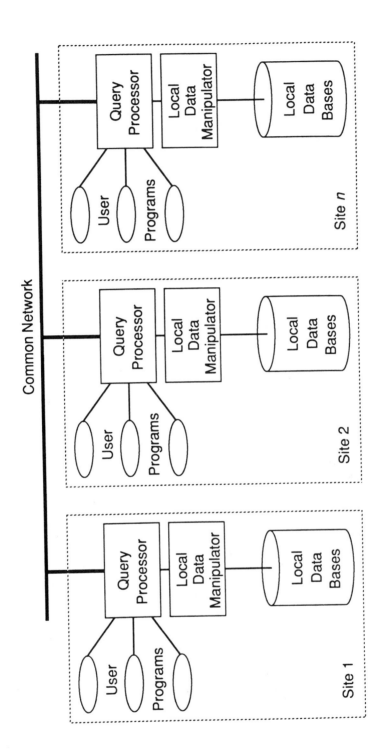

Figure 9. Distributed Data Administration Architecture

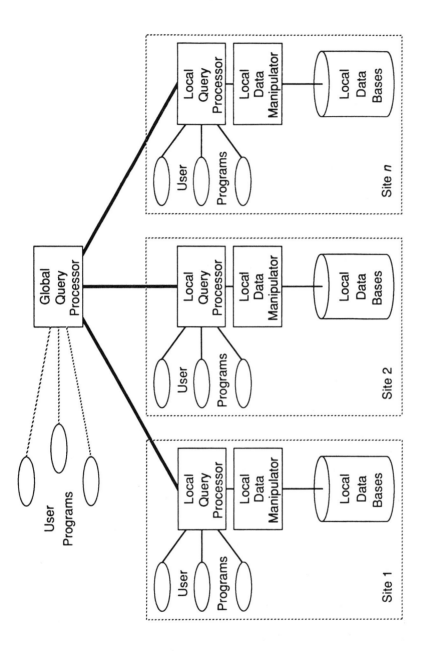

Figure 10. Hybrid Data Administration Architecture

273

administration system itself. There are a number of ways of implementing such an architecture, but they are all characterized by standardized interfaces between the local data systems and the global query processor.

C. IMDAS - The AMRF Data System

The AMRF data system is called IMDAS - Integrated Manufacturing Data Administration System. Much of what follows has been excerpted from [8] and [9]. IMDAS manages many different databases (see Figure 11). It uses the Semantic Association Model (SAM*) as the integrating model [10]. The resulting model is a semantic network which is capable of modeling the complex objects, relationships, and constraints typical of most manufacturing systems. Users make requests from IMDAS using a data manipulation language (DML) which resembles SQL, the existing ANSI standard.

The IMDAS administration architecture is a hybrid architecture containing three classes of service modules - BDAS, DDAS, and MDAS - as shown in Figure 12. The following sections describe these modules as they were originally designed.

1. Basic Data Administration System (BDAS)

The BDAS (Basic Data Administration System) provides a uniform interface between the data residing on each component system and the integrated database. There is therefore a BDAS software module resident on every component system in the automated factory. Since the component computer systems have widely different processing capabilities, the BDAS provides only the essential data management and communication functions. These functions may be implemented differently in each component system to support the unique hardware and software environment of the system. The functions of the BDAS are interprocess communication, network communication, data translation, command translation, and data management (see Figure 13).

a. Interprocess Communication: If the component system houses multiple control and sensory processes, or if the communication and data management services are implemented as separate processes, a mechanism for communication among these processes is needed. The AMRF uses a shared memory concept which permits data to be used by more than one process without explicit action by the originator to deliver it to all users. The shared memory can be considered as a database in its own right, as well as a data communication vehicle between the control processes and the data services. The shared memory approach is particularly effective in equipment level systems which must perform real-time data acquisition (see section IV.E.1).

b. Network Communication: Every component system must have access to the factory data network and must implement the protocols necessary to communicate with the other component systems. The IMDAS architecture

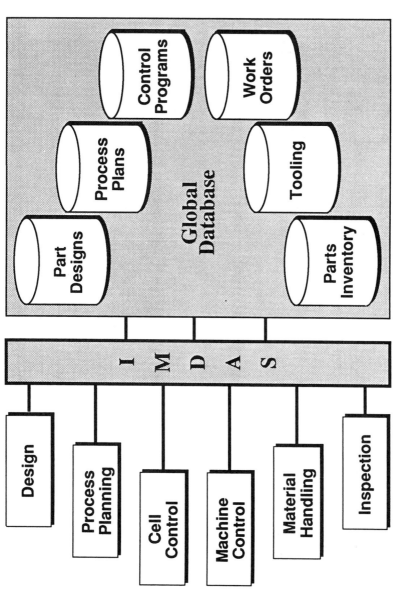

Figure 11. IMDAS Distributed Databases

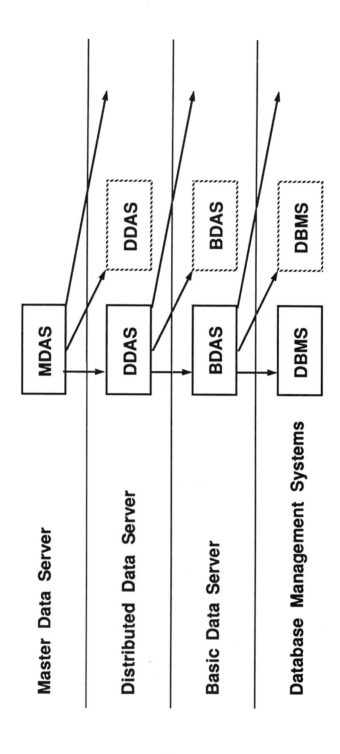

Figure 12. IMDAS Data Administration Architecture

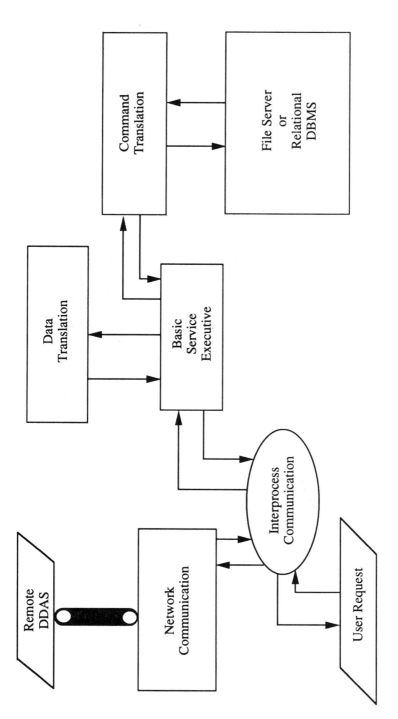

Figure 13. Functional View of a BDAS

assumes a peer-to-peer relationship between any two component systems. The four lowest layers of the ISO/OSI reference model (see section IV.A) must be implemented for each component system. These basic protocols allow reliable communication between any two systems on the network. IMDAS has defined its own layer 7 protocols to implement distributed data services.

c. Data Translation: Whenever data are to be moved from one system to another, it is necessary to translate the data from the source representation to the required target representation. The data translation process typically involves the translation of data type, structure, and format. This translation can be performed by either direct source-to-target conversion or source- to-common-to-target conversion. IMDAS uses the latter, since it requires substantially fewer translators at each component system: one local-to-common and one common-to-local. Using this approach, the data translator converts the data from the local representation to an "interchange" representation on "outbound" sessions and from the "interchange" representation to the local representation on "inbound" sessions. The interchange format strongly resembles the STEP physical file structure, but the data encoding follows the rules set out in ASN.1 (Abstract Syntax Notation) standard.

d. Command Translation: Transactions are issued to a BDAS in the form of a query tree of standard primitives. The function of command translation is to take each operation in standard form and translate it into the query language or access mechanism understood by the underlying physical data management tool, which may be a data manager or file manager. Command translation can often be table driven although some particularly complex data structures may require preprogrammed procedures.

e. Data Management: The BDAS in each component system must provide, through the local database or file management system, the actual access to and manipulation of the local databases. The database manipulation and management capabilities can vary substantially from one system to another, depending on the facilities provided by the software. In small control systems, the local database may be implemented only in the main memory and managed by a shared memory management process. In others, a file manager may meet all local requirements, giving the system little more than "read" and "write" operations. In larger systems, sophisticated database management systems may provide the actual database services. Moreover, there may be more than one such facility on a single component system.

The sequence of primitive operation commands given to a BDAS command translator must be within the capabilities of the underlying data management system which manages the data. The data manager must also be capable of cooperating with the other data managers in controlling and maintaining consistency, concurrency, and recovery.

f. Basic Service Executive: This software module provides a single point of contact between the DDAS and the BDAS data services. The basic service executive accepts DDAS commands, routing them to the BDAS modules which perform the command translation, data manipulation, and data translation functions. It also constructs local data delivery paths, sequences the execution of query tree operations and reports the completion or error status back to the DDAS transaction manager.

The Basic Service Executive, because it communicates with the DDAS and with its peer BSEs must have a standard set of interfaces. The interfaces to the subordinate BDAS modules are to some extent determined by the local system conventions.

2. Distributed Data Administration System (DDAS)

The DDAS provides control and user processes (its users) with uniform access to the global database. Each DDAS is responsible for providing data management services to every process within the component systems assigned to it. It further provides and controls the distributed database management functions for all the data maintained by the BDASs on these component systems. The major modules originally included in the DDAS are: the distributed service executive, the data manipulation language service, the query mapping service, the transaction manager, data directory services, and the data assembly process. Figure 14 illustrates the interrelationship among these modules.

a. Distributed Service Executive: This module oversees all DDAS activities and provides the single point of contact between the data system and some set of users and between the DDAS and the MDAS. It is responsible for initialization of the DDAS and its subordinate BDASs and for coordination of all DDAS functions.

User processes running on the DDAS component systems issue data management transactions (often called "queries," even though they may be update requests) through the local or network interprocess communication services. Upon receipt of a transaction from a user process, the DSE routes the transaction to the Data Manipulation Language Service, and then determines whether the transaction can be locally executed. If it can, the DSE passes the transaction to the Query Mapping Service and then to the Transaction Manager for execution. If the transaction cannot be locally executed, the DSE sends it to the MDAS for service. When execution of the transaction is complete, the DSE reports the completion status to the originating user. It performs the same functions for transactions originating from other DDAS domains which are passed to this DDAS by the MDAS Transaction Manager.

b. Data Manipulation Language Service: The DDAS receives transactions expressed in the IMDAS DML from the user or control processes within its domain. A DML query is issued against the global external view of the integrated database by the user process. The DML Server parses the DML

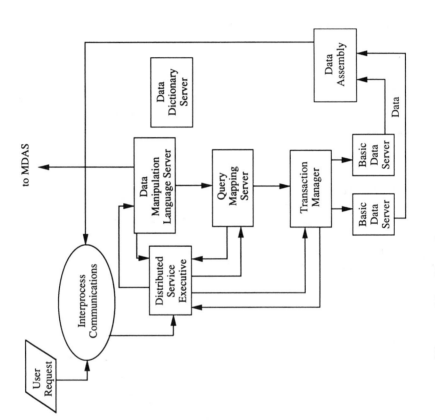

Figure 14. Functional View of a DDAS

query into a tree of primitive operations on the external view and then modifies the query tree to incorporate the mapping from the external view to the global conceptual view. In addition, the DML server encodes into the query tree data delivery operations, which reference the user-designated source or destination data areas. The query tree may be then further modified to incorporate security and integrity constraints defined as part of the global conceptual view. The modified query tree now constitutes a transaction on the global conceptual view, representing the user's request.

The DML Server then identifies whether the requested operations can be performed entirely by this DDAS. If any of the data referenced are out of this DDAS domain, the query in its translated form is flagged for transmission to the MDAS for further processing. Otherwise, the query is flagged for processing by the local Query Mapping Service.

c. Query Mapping Service (QMS): This service accepts query trees which can be satisfied entirely within the sub-databases of this DDAS. These queries may have come from local users or from remote users via the MDAS. The Query Mapping Service decomposes each query into one or more subqueries on the fragmented views reflecting the partition of data across subordinate BDAS databases and the capabilities of the data servers managing those databases. At this time, integrity constraints defined in the mapping from the global conceptual view to the fragmented views are incorporated into the query tree. The resulting query tree now constitutes a collection of transactions on individual databases which together 1) represents the original conceptual operation, and 2) will guarantee that the distributed database will be in a consistent state. The query tree is then restructured and optimized so that the query tree consists of an optimum sequence of operations. The optimized query tree is organized into subquery trees, each of which can be executed by a single subordinate BDAS command translator. The intermediate and final data assembly of BDAS results constitutes another subtree, which is assigned to a data assembly service somewhere in the DDAS domain. The subquery trees, including intermediate and final delivery operations, and precedence relationships, are now passed to the transaction manager for execution.

d. Transaction Manager (TM): The transaction manager performs the control and management of distributed query execution, including enforcement of database integrity, concurrency control and recovery. The integrity control mechanism of the TM realizes a serial execution schedule for each transaction received from the Query Mapping Server, while ensuring that the transaction is processed as an atomic unit of recovery. The recovery facility of the TM brings the database back to a consistent state after a system failure or a violation of the integrity constraints has been detected. The concurrency control mechanism detects conflicts between active and pending transactions by identification of common attribute references. Pending transactions which would conflict with active transactions are deferred, held in the TM and released when the conflicting transaction completes.

When a transaction is released, the TM transmits the proper sub-trees of the query tree, along with the directions for constructing delivery paths, to the proper BDAS. The TM oversees the distributed execution of these operations, receiving status reports from all of the BDASs involved. When the final results of the transaction have been delivered to the requesting process or when the data is properly updated, the TM identifies the transaction as completed.

e. Data Directory Service (DDS): Whenever a DDAS is brought up, it must initialize its data directory to reflect only the data which is resident in the local BDAS. It then creates connections to its subordinate remote BDASs, requests and incorporates their data directory information. The result of the integration is a mapping of a segment of the global, conceptual view onto the entire sub-database constituting the domain of the DDAS. This view represents a comprised view of the BDAS sub-databases, in which all conflicts and inconsistencies existing in the BDAS directories are resolved. The DSE initializes the interfaces to the control processes served by the DDAS and is then ready to begin accepting requests for service.

The DDAS data directory is built from the directory information provided by the subordinate BDASs when they are integrated. It includes the following metadata:

1) a complete representation of the global conceptual model

2) the "mapping schemas" relating each BDAS repository to the global conceptual model

3) the integrity constraints and other relationships associated with the data in subordinate BDASs,

4) the mapping information for the external and conceptual views of the data referenced by the control processes which this DDAS supports,

5) the security constraints associated with the views of the conceptual database referenced by the local control processes,

6) the information representing the capability of each subordinate DBMS or Command Translator.

When requested by the DML server or the QMS or the TM, the Data Directory Server performs a metadata lookup and provides data location, structure, and delivery information.

f. Data Assembly Service (DAS): This is a virtual module which performs selection, join, union, intersection, merging, and sorting of data retrieved from subordinate databases, and multiple projections of user-provided data for distribution to subordinate databases for updates. This function is required for transactions in which multiple databases are involved and for transactions beyond the capabilities of a single repository.   While this was originally conceived as a DDAS function, it can, in principal, be performed by any BDAS repository that can accept dynamic schema additions and load data from external sources.

3. Master Data Administration System (MDAS)

The function of the MDAS (see Figure 15) is to coordinate the activities of multiple DDASs. This coordination consists primarily of managing the master data directory, resolving concurrency problems among DDASs, and directing initialization, integration and recovery. The software needed to accomplish this function is resident in every DDAS.   Because it is largely identical to the software used to integrate BDASs, MDAS software modules differ in the construction and referencing of a master directory containing the data definition of the global conceptual view and the fragmentation mappings to the individual DDASs. Therefore, the MDAS represents another instantiation of the DDAS software, which is comprised of a query mapping service, a transaction manager, and a data directory service. The unique MDAS software element is the Master Service Executive, which oversees global integration, reconciliation and recovery.

The Master Service Executive treats the DDASs as its users, receiving transactions in the form of query trees referencing the global conceptual view, instead of DML referencing user external views.   It routes them to its QMS, which maps the global query tree onto a set of fragmented views, each representing that portion of the global conceptual database maintained by a single DDAS. The MDAS QMS is unaware of the partitioning and mapping of this database into the sub-databases supported by the DDASs.   It sees the DDASs as a set of uniform subordinates, each of which has the full spectrum of data management capabilities including assembly, and each of which manages a fragment of the global conceptual database.   The fragmented-view queries delivered by the QMS are then sent to the Master Transaction Manager, which distributes the fragments to the appropriate DDASs and oversees the execution of the query.   When the transaction is completed, the MDAS TM reports the completion to the MSE. The MSE sends the report to the originating DDAS, who is responsible for the communication with the originating user.   As in the case of all DDAS operated queries, the data delivery operations are contained in the query tree and the final data delivery is direct - from the system which executes the root operation in the query tree to the designated area on the user's system.

While the Master can use the conflict detection mechanism employed by the DDAS TMs, this mechanism, which successfully avoids concurrency problems at the BDAS level, cannot guarantee the avoidance of concurrency problems at the DDAS level, because the DDAS may be executing transactions of local origin, about which the MDAS has no information. For this reason, the MDAS must employ a cooperative concurrency control protocol on every transaction which is fragmented to more than one DDAS.

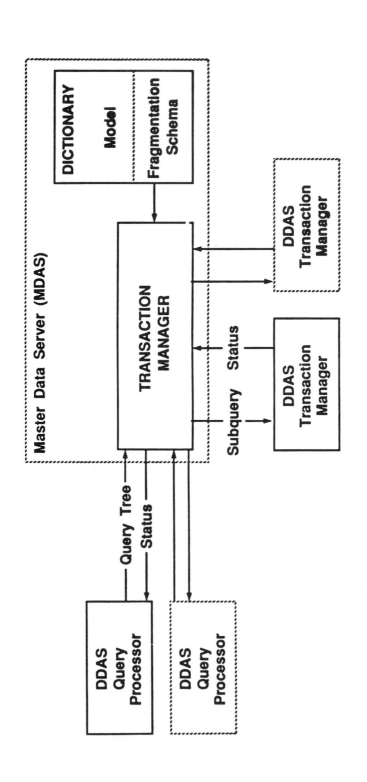

Figure 15. Functional View of an MDAS

In the integrated factory, there must be exactly one MDAS. When the data system is initialized, an operator designates one DDAS as the MDAS and directs it to integrate the other DDASs. The DDAS services at the MDAS site continue to function as for any other DDAS. If an MDAS goes down, an operator designates one of the remaining DDASs in the partition as the new MDAS. The new MDAS must rebuild the master directory and initiate a recovery procedure. When the data system is partitioned to allow a number of subnetworks to run independently, an MDAS is established for each partition containing more than one DDAS. Each MDAS requests its subordinate DDASs to provide their directories and integrates them to form its own master directory. When the sub-networks are subsequently reintegrated into a single data system, the MDAS uses the recorded information to reconcile the sub-databases.

4. Performance

Performance tests were carried out on IMDAS and reported in [11]. The following is excerpted from that report.

From the point-of-view of operational characteristics, IMDAS transactions issued by AMRF controllers can be broken down into seven categories:

1) creation of new entries in the form of large bodies of simply structured information such as process plans or NC programs.

2) retrieval of large bodies of simply structured information by a few key values.

3) Retrieval of small simple relations of work-in-process data such as tray definition, lot description, etc.

4) Update of simple relations of work-in-process data.

5) Update of lots and kitting orders by insertion of rows.

6) Retrieval of small reports which require complex manipulation of the underlying databases - production of views very different from the storage organizations.

7) Update of complex relations requiring spawning of secondary update transactions.

Categories (2), (3) and (4) represent over 80% of the actual transactions issued by controllers during "production runs." Categories (6) and (7) represent the remainder of the production load. The insertions in category (1) are uploads of engineering data developed on various systems and occur sporadically during production.

In 1987 (see Figure 16) response times of 30 seconds or more for transactions in the first five categories were typical. Furthermore, response times of several minutes for transactions in the latter two categories were not uncommon. This was determined to be partly a consequence of the quality of the prototype IMDAS implementation and partly a consequence of the total load on the system.

By 1989 (see Figure 17) IMDAS response times had improved considerably. The 1989 response times are, by category, as follows:

1) Insertion of new process-plans, etc.: 5-6 seconds.
2) Retrieval of process-plans, etc.: 5 seconds.
3) Retrieval of simple relations: 5-10 seconds.
4) Update in simple relations: 5-10 seconds.
5) Update by insertion: 5-10 seconds, depending very little on the number of new entries.
6) Retrieval of views requiring complex manipulations: 10-45 seconds, depending on the complexity of the operation. These vary considerably.
7) Update which spawns secondary updates: 30-45 seconds.

The first five categories represent essentially simple database manipulations, even though some are retrievals and some are updates. The variation in performance depends largely on which actual database contains the information. By comparison, the latter two categories represent relatively complex data manipulations which may require several separate operations on a relational database and can therefore be expected to take longer.

While these response times are marginally acceptable in the AMRF, it is clear that without considerable improvement in performance, IMDAS could not be used in a real production facility. Since IMDAS is a prototype, production quality performance has not thus far been a goal of the project. The question then is: To what extent can IMDAS performance be enhanced by improvements in the implementation rather than changes in the design?

We can break down the time consumed by IMDAS for a given transaction into four elements:

1) Propagation time - the time required to get the transaction from the application process (controller) to the IMDAS and through the IMDAS to the affected Command Translator modules (CTs), plus the time required to get the completion status from the CTs through the IMDAS hierarchy and back to the application.

2) Transaction analysis time - the time required by the DDAS to interpret the transaction string into the complete operation tree in "query interchange form" directed to the proper Command Translators.

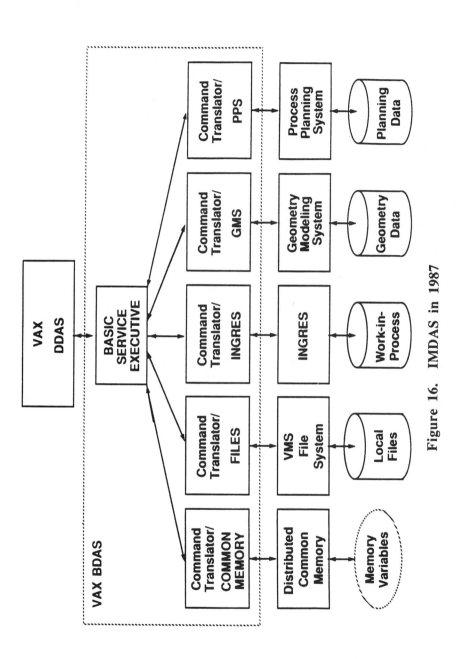

Figure 16. IMDAS in 1987

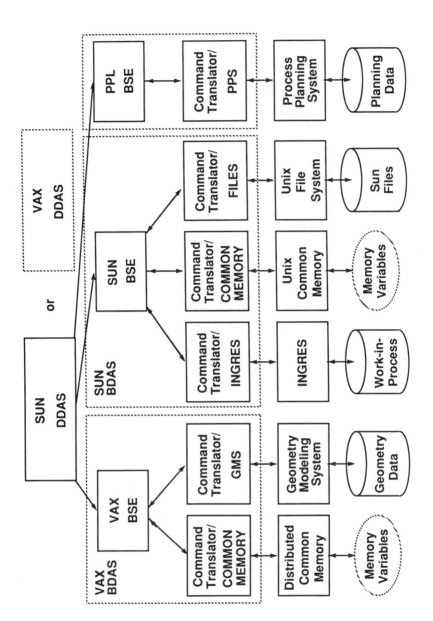

Figure 17. IMDAS in 1989

3) Execution time - the time required by the CT and the underlying database management system to execute the transaction and convert the data to/from interchange form.

4) Editing time - the time required to convert the input or output data between the interchange form and the report form in the user data areas.

The following table gives a rough breakdown of the time, in seconds, required to execute various transactions.

Table II. IMDAS Timing Data

Timing Element	Simple Insertion or Retrieval	Complex transactions on one database	Transactions with multiple subparts
Propagation	2.5 sec	2.5 sec	up to 4 sec
Analysis	1.2 sec	1.5 sec	up to 2 sec
Execution	0.5-4.0 sec	10-40 sec	25-40 sec
Editing	< 0.5 sec	< 0.5 sec	< 0.5 sec

Many of these delays can be attributed to implementation choices rather than design choices. Consequently, dramatic improvements are expected as modules are rewritten. In addition, we expect to change the "hierarchical view" of the data to an "object-oriented" view which will also improve the performance.

IV. ISSUES IN COMMUNICATIONS

The AMRF communications system provides those functions needed to transmit information between computer programs executing production and data management tasks. In planning an integrated manufacturing network, we believe that three ideas are fundamental:

1) that the production management and data management programs themselves use one common connection service specification for communication with other programs, regardless of function or location;

2) that the physical networks are transparently interconnected, so that any program could conceivably communicate with any related program anywhere in the complex;

3) that the technology and topology of subnetworks are chosen to provide optimal communications responsiveness for the primary functions.

In this section we describe a communications architecture whose design is based on these ideas. This architecture ensures that ANY production management or data management architecture that is deemed to be desirable can be conveniently constructed with the network as-built. We also describe the actual AMRF implementation of these ideas.

A. Types of Communication

Communications can be divided into two classes: those WITHIN computer systems, and those ACROSS computer systems. The first type is often referred to as "interprocess communication" while the second is often called "network communication."

Interprocess communication is dependent on features of the operating system. Many systems provide no such facility at all, or provide only for communication between a "parent" process and "child" subprocesses which are created by the parent. On such systems the coordination of multiple production management and data management activities is extremely difficult. On the other hand, properly implemented "network communication" software provides for the case in which the selected correspondent process is resident on the same computer system as the process originating the connection. That is, the proper solution for the future is to make local interprocess communication a special case handled by the network software. This solution has the added advantage that all communications by a production management or data management process, regardless of the location of the correspondent process, has the SAME interface.

The accepted paradigm for network communication is the Open Systems Interconnection Reference Model (OSI) which separates the concerns of communication into the following 7 layers [12].

1) Physical layer deals with cables, connectors and signals, and the protocols governing "who talks when" on a shared medium.

2) DataLink layer deals with packaging the signals into elementary messages (called frames), checking for errors in transmission and (perhaps) recovering lost frames. It provides the control and checking for the physical link.

3) Network layer deals with making logical end-to-end connections out of one or more physical connections; i.e., finding a path that gets a message from station A to station B.

4) Transport layer controls and checks the end-to-end connections so that complete messages are delivered in logical order and without losses.

5) Session layer distinguishes separate processes or functions communicating between the same two stations, and implements rules for message-flow between those processes or functions.

6) Presentation layer converts between local data representations and interchange data representations.

7) Application layer deals with establishing links between processes and the relationships between interprocess links and functions being performed.

B. Network Architecture

A great deal of flexibility is created by implementing the OSI model. On one hand, a single physical medium can multiplex many separate process-to-process communications. On the other hand, a given process-to-process connection can use several separate physical connections with relays between them. This gives rise to the general manufacturing network architecture shown in Figure 18.

Ideally, all stations on the network implement common OSI protocol suites in the intermediate layers (3-5) and some globally common protocols for moving data sets in layers 6 and 7. In addition, other standard application layer protocols will be shared among systems performing related functions. The choices of protocol suites in the Physical and DataLink layers and the connectivity of individual stations will vary. They will depend on the physical arrangement and capabilities of the individual stations, and their functional assignments and performance requirements. There may be one physical network, or many. All of these separate physical networks, however, must be linked together by "bridges" that implement the proper Network layer protocols. This results in a SINGLE LOGICAL network on which any given production management or data management process can connect to other process regardless of location. We note, that because this architecture is layered, multiple subnetworks become transparent to our interprocess communication paradigm.

The "enterprise networking" concept, connecting Manufacturing Automation Protocol (MAP) networks with Technical Office Protocols (TOP) engineering networks [13], demonstrates that the generalized network architecture is, in fact, currently practical. We believe that this is the appropriate network architecture for manufacturing. It is likely that emerging physical networking technologies will, in time, make the physical layer standards selected by MAP/TOP obsolete. This will lead to the addition or substitution of subnetworks with new physical and datalink protocols to current networks. Nevertheless, the transparent, multiple, subnetwork architecture shown in Figures 18 and 19 should result in little or no impact on networks already in place and on process-to-process communication. At the same time, adherence to at least the layering, but preferably also the intermediate layer protocol suites, in various types of "gateway" machines, provides for the transparent interconnection of subnetworks based on proprietary, or nonstandard protocol suites in the lower layers.

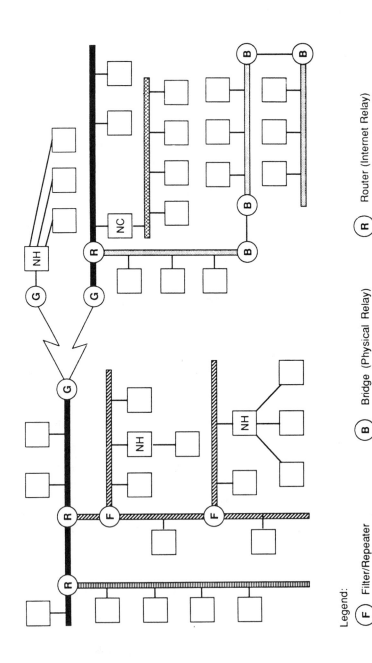

Legend:

(F) Filter/Repeater

(G) Gateway (Message Relay or Protocol Converter)

(B) Bridge (Physical Relay)

NH Networked Host (Attachment point for others)

NH

(R) Router (Internet Relay)

NC (Sub)Network Controller

Figure 18. Generalized Manufacturing Network

292

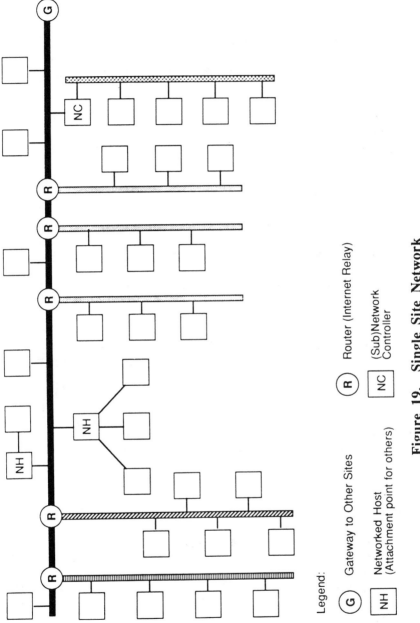

Legend:

(G) Gateway to Other Sites

(R) Router (Internet Relay)

NH Networked Host
(Attachment point for others)

NC (Sub)Network
Controller

Figure 19.   Single Site Network

C. Technology

There are now many standard protocol suites for the DataLink and Physical layers, and there will soon be more. They all provide frame delivery and integrity checking. Some provide for reliability and recovery. Others defer those considerations to the transport layer. The real distinguishing characteristics among these standards are the signalling technologies and the sharing algorithms. Loosely speaking, the signalling technology determines the raw transmission speed, the relative immunity to electronic noise, and the cost. The sharing algorithm determines the nature of network service seen by the station. There are generally three choices:

1) connection to one other station or one other station at a time, with fixed dedicated bandwidth (point-to-point, time- and frequency-division);

2) connection to multiple stations simultaneously, with variable bandwidth with fixed lower and upper bounds depending on the number of stations connected to the medium (token bus and ring);

3) connection to multiple station simultaneously, with variable bandwidth from zero to the bandwidth of the medium depending on the traffic generated by all stations connected to the medium (CSMA/CD).

In general, engineering and administrative activities, which have infrequent and variable communications requirements, can tolerate and use the type (3) services more effectively. The production control activities, which have frequent and regular messaging requirements, however, prefer type (1) or (2) services.

There are also several "standard" protocol suites for the intermediate layers as well. But in this area, the differences are historical rather than functional. It is clear that OSI existing intermediate layer protocols will be THE standard in the near future. In the upper layers, standards are still evolving. Here the only problem will be to determine the suite of protocols necessary to a given production management or data management function.

D. Topology

Topology is that aspect of network design which concerns itself with the connection of stations to subnetworks and the interconnection of the subnetworks. Topology is at least as important as bandwidth and access protocols in determining the effective performance of integrated networks. Processes which need to communicate frequently should be directly connected, or connected to a common bus/ring, if at all possible. On the other hand, only

two factors should really motivate dividing a network into subnetworks:
1) the feasibility or cost of connecting all of the potential stations to
   the same physical network;

2) the ability of the single network to carry the total traffic load.

Several varieties of bus/ring networks have limitations on the total number of stations which can be connected, or on the total cable length. When this limit is reached, partitioning is unavoidable. In addition, the performance of most bus and ring networks is inversely proportional to the number of stations or volume of messages placed on the network. When the performance of a subnetwork degrades the performance of the primary production management or data management functions using it, it is time to partition that subnetwork or replace the networking technology. The former is usually adequate, easier and cheaper, and adherence to the OSI model should make it invisible to the communicating processes.

Ideally, the generalized manufacturing network architecture in Figure 18 will be implemented in the much more restricted form depicted in Figure 19. There is a common "spine" or "backbone network" which connects to ALL subnetworks, although some the of the subnetworks may be directly interconnected. This architecture guarantees that the maximum number of relays on any process-to-process connection is two. While this is not always practicable, it is, in our view, always the desirable goal for the network architecture of a single site.

E. The AMRF Communication System Circa 1988

In the following sections, we describe how these ideas were implemented in the AMRF. Much of the remaining sections have been excerpted from [14].

1. Common Memory

AMRF researchers realized that, in the "factory of the future," manufacturing functions (regardless of the level of automation) would run on many different computers, of all sizes and models, and possibly located in different buildings. The major implication of such a "distributed" system is that information transfer would have to be accurate, reliable, and independent of the actual physical location of the machines. In addition, these transfers would have to be accomplished without interfering with or adversely affecting real-time processes on the receiving machine. The AMRF solution was, and continues to be, based on a global common memory.

Originally, common memory was developed to support real-time data analysis and robot control in a multiprocessor configuration using a physical common memory. Each of these processors was a single board computer. They were all connected to a shared-memory area that was on a separate board. This area maps into the address space of each processor, allowing it to read and write to this "common memory area" and thereby communicate with the other processors. The advantage of this approach is that the sender does not have to wait for the receiver; he can send whenever he is ready. The receiver will read messages whenever he is ready. The disadvantage is that important messages can be lost. Arrangements must be made in advance so that this is avoided.

In 1981, we began the integration of many pieces of equipment. It was obvious that a single physical common memory was no longer possible or practical. Therefore, each computer system had its own local common memory. These common memories were connected through locally-developed network services that are transparent to the user process. The result was a global logical common memory.

The AMRF common memory uses the concept of computer "mailboxes". Mailboxes are logical storage areas where messages (called "mailgrams") are placed by the sender process and picked up by receiver processes. The mailgram transfers across computer systems are accomplished by the communications systems in a fashion totally invisible to the sender and receiver processes. Moreover, the communications system operates asynchronously. This means that application processes can leave "messages" for each other and stop to read their own "mail" at opportune times without interrupting each other. The common memory mailbox implementation permits data to be used by more than one process without explicit action by the originator to deliver it to all users. There are three cases to consider.

First, the retrieving process resides on the same computer system as the originating process and has direct access to the same memory address area. In such cases, it can "read" the information unit simply by fetching the mailgram from the common memory space. In Figure 20, the arrows show the flow of command and status information: application process 1 deposits data into common memory mailbox A, which is read by both application processes 2 and 3. Process 2 generates a message that is deposited into mailbox B for process 3 to read, based on the data in mailbox A.

Second, the retrieving process may reside in the same computer chassis (i.e., a Multibus system) as the originating process, but have no direct access to the originator's memory. In such cases, a transporting process resident on a separate processor board must copy the mailgram from the originator's mailbox to an intermediate location then to a mailbox designated to receive that information in the local memory area of the retriever.

Finally, in the case that the retriever resides on a physically separate computer, there is a Network Interface Process (NIP) resident on the originator's and the retriever's node. The originator's NIP uses the local shared memory protocol to read the originator's mailgram and transmits a copy over the AMRF network to the NIP on the retriever's system. The receiving NIP then stores the mailgram into the appropriate mailbox on that system, where it can be read by the retriever using the local protocol there.

Figure 21 displays the distributed common memory. As in Figure 20, information is passed between three processes. However, because these three processes are located on three different (remote) processors, the NIP passes the information through the network and into and out of the appropriate common

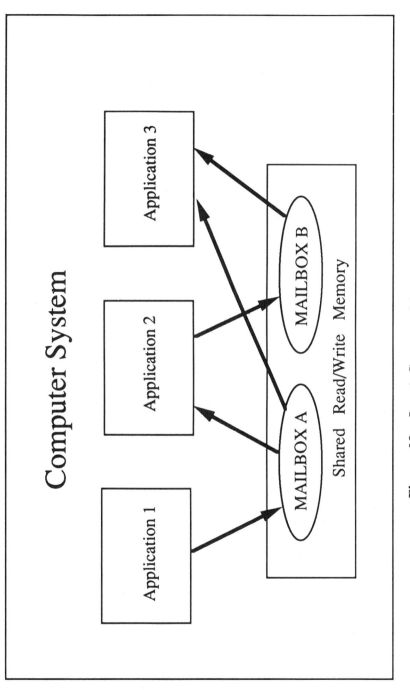

Figure 20. Local Common Memory

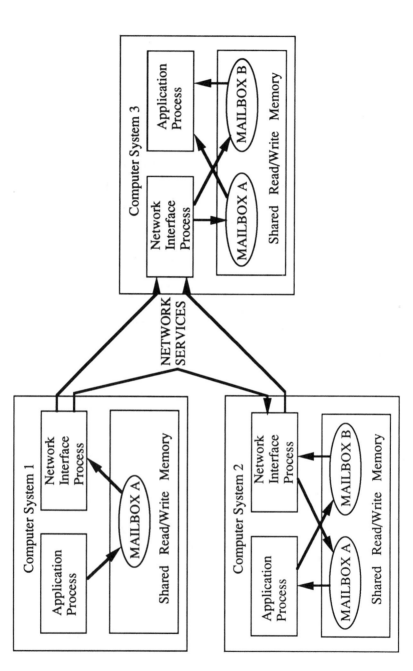

Figure 21. Distributed Common Memory

memory mailboxes. The arrows show the flow of command and status information.

2. The AMRF Network

At the time of inception of the AMRF (1981), there were no national standards by which machines and controllers of various manufacturers could be expected to intercommunicate. Consequently, many of the decisions made during that time would be inappropriate today. The AMRF network is based on the OSI architecture described in section IV.A.

Beginning in 1982, the AMRF sought to select then-emerging standards for the protocols in the layers of the AMRF network. Some of the AMRF standards choices fortuitously coincided with the MAP choices; others, largely owing to the availability of limited choices in 1982-4, did not. The currently operating AMRF network is largely a collection of interim protocols which are intended to be gradually supplanted by commercially available nonproprietary network protocols, as individual component systems were upgraded or fit into the whole complex.

a. Link and Physical Layers: We paired the Link and Physical layers because in many cases they were paired in the available products, and because the choice of data link protocol depended on the characteristics of the underlying physical protocol. This is the first and most significant area in which the AMRF network requires the support of alternative protocols. The original physical protocol selected was EIA RS232C full-duplex, asynchronous, point-to-point modemless connection at 9600-baud.

The associated link layer protocol is AMRF-originated, providing frame definition and integrity checking only. This technique provides for frame definition and integrity checking with the "best effort" philosophy: erroneous frames and lost frames are discarded entirely by the link layer service, so that only correct frames reach the receiving network layer. This is analogous to the data link protocols employed by the IEEE 802 standards. It assumes that availability of transport layer protocols to recover and retransmit lost information units.

Later, IEEE 802.3 bus networks were added as subnets to connect many facility components. Finally, a MAP token bus network was installed as the backbone network.

b. Network Layer: The ISO 8473 Connectionless Network Service Protocol was specified to provide for host-to-host message routing services. Briefly, this is a "datagram" protocol, in which each data unit is labelled with the sending and receiving host and finds its way through the network from the sender to the receiver without regard to any previous or concurrent transmissions.

This permits the introduction of "internet gateways" (what MAP calls "routers") - stations on two or more networks which receive messages on one of the networks and retransmit them over another of the networks toward the destination indicated in the network layer envelope. Since the AMRF network is intrinsically a multi-network involving several separate physical protocols (with associated data link protocols), any of the hosts can be connected to more than one network. If such a host receives a data unit for which it is not the indicated destination, it can function as a gateway by retransmitting the data unit by whatever means it would have used to reach the indicated destination. In addition, "gateway stations" have been procured to link the major subnetworks to the token bus.

Because the destination is clearly specified in the data unit, a receiving station can determine whether the data unit is intended for that station or must be relayed to another; and because the source is clearly specified, the destination station can determine the true originating host regardless of the path over which the data unit arrived. To make this simple mechanism work globally, the AMRF network specifies the network layer protocol as mandatory, even when a point-to-point link or direct connection to a common bus is used. The network layer software is then required to support multiple underlying physical/link protocols (most stations have serial connections and bus connections) but not to understand any qualitative differences between them.

c. Transport Layer: In the area of transport protocols, the AMRF network is in a state of change. The nominal standard identified in 1984 - the ISO standard, which coincided with the MAP choice - had no commercial implementations available until 1986. This necessitated the implementation of an interim transport service protocol for the engineering phase of the AMRF, which was then used from 1983 through 1986. Moreover, the AMRF network architecture anticipated the use of a single global transport protocol, which made it difficult to accommodate adoption of the standard on some stations while it was still unavailable on others.

The final goal is the ISO 8073 Transport Layer Service Protocol Class 4. The class distinctions restrict transport services, and thereby complexity, according to the degree of simplicity and reliability provided by the lower layers. The AMRF multi-network environment mandates class 4 services - the highest class, which assumes little reliability in the lower layers - but, because it is a local area network, very few of the extended options. This is essentially identical to the MAP transport protocol selection.

The interim standard is the Transmission Control Protocol for Defense Networks (TCP), MilSpec-1778. This protocol is not strictly a "transport" protocol in the OSI model sense, since, in addition to transport functions, it contains a limited session management protocol and a primitive application selection protocol (which used to be thought of as a session-layer function) as well.

This standard is used in the AMRF network only on those hosts which use this protocol as part of a large class of distributed services offered by the manufacturer and which do not support any other protocol (simultaneously) over the principal network. Use of this protocol, and thus the principal network, affords efficient communication among these stations.

d. Session, Presentation, and Application Layers: The AMRF network has a nominally "void" session layer. Unlike the few existing ISO application layer standards, the operating AMRF application layer service (common-memory mapping) is a station-to-station service which is itself a multiplexer. That is, the single mapping-service to mapping-service connection may (unwittingly) carry any number of logically separate communications. Thus the "session" layer is subsumed.

The AMRF network currently has a "void" presentation layer. All control interchanges are in some form agreed to by the parties involved. It is expected that the incorporation of the distributed data protocols in a future version of the network architecture will result in the standardization of some interchange form for all message units, which may obviate presentation layer protocols indefinitely.

The AMRF network currently provides only one "application service" at all stations: the "memory mapping service." The memory mapping service is the means by which the "common-memory" concept is extended to multiple computer systems.

Figure 22 shows the topology of the 1986 AMRF network. The network is primarily comprised of point-to-point connections using serial, RS232 connections. Additional network pathways were provided using Ethernet (TCP/IP) to accommodate greater traffic loads while simultaneously providing enhanced speed. Two subnets were also included. Figure 23 shows the 1988 configuration in which several of the aforementioned network standards were incorporated.

e. Remarks: As noted above, many of these decisions were made before many of the current international standards were in place. The AMRF continues to integrate new standards such as the Manufacturing Messaging Specification as they become available.

V. SUMMARY

In this chapter, we have discussed several integration issues in the AMRF. In particular, we examined the AMRF system architecture. We have argued that such an architecture has three separate but interrelated management components: production, information, and communications. We described the architectures themselves, some of the principles employed, and experiences gained from implementing those architectures.

The AMRF was the first system to implement separate architectures for production, information, and communications management. As such, it was a turning point in the design of automated systems. The architectures described in the preceding sections have provided the foundation for similar efforts in many universities and companies in the US and throughout the world. In addition, they have had, and will continue to have, an important impact on the development of international standards in areas related to all aspects of manufacturing. Finally, they form the basis for all ongoing work in the AMRF.

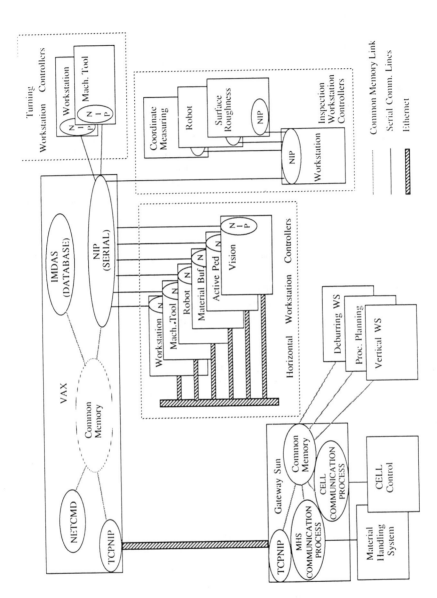

Figure 22. Topology of 1986 AMRF Network

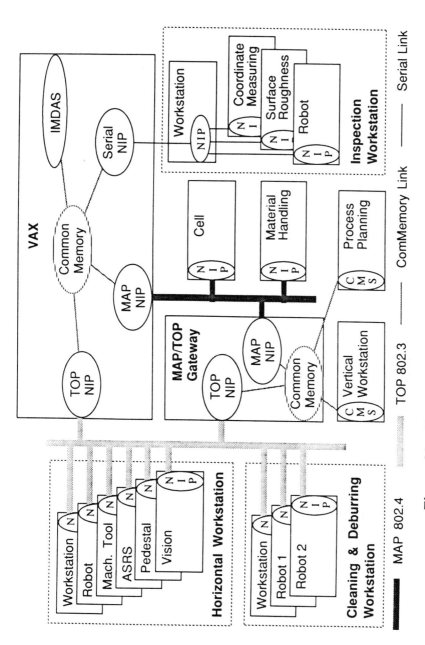

Figure 23. Topology of the 1988 AMRF Network

ACKNOWLEDGEMENT

The author wishes to thank Edward Barkmeyer for his contributions to this chapter. Many of sections were excerpted from papers he has written and his editorial comments have improved the chapter immeasurably.

REFERENCES

1.   J. Simpson, R. Hocken, and J. Albus, "The Automated Manufacturing Research Facility of the National Bureau of Standards," *Journal of Manufacturing Systems,* **1, 1,** 17-31, (1982).

2.   B. Smith and G. Rinaudot, "Product Data Exchange Specification: First Working Draft," NISTIR 88-4004, 1988.

3.   B. Catron and S. Ray, "ALPS - A Language for Process Specification," *International Journal of Computer Integrated Manufacturing,* (to appear).

4.   J. Albus, A. Barbera and R. Nagel, "Theory and Practice of Hierarchical Control," Proceedings of 23rd IEEE *Computer Society Conference,* 18-39, 1981.

5.   A. Jones and C. McLean, "A Production Control Module for the AMRF," *Proceedings of the International Symposium on Computers in Engineering,* 1985.

6.   R. Jackson and A. Jones, "An Architecture for Decision-Making in the Factory of the Future," *INTERFACES,* **17,** 15-28, 1986.

7.   W. Davis and A. Jones, "A Generic Architecture for Intelligent Control Systems," NISTIR 91-4521, 1991.

8.   E. Barkmeyer, M. Mitchell, K. Mikkilineni, S. Su, and H. Lam, "An Integrated Manufacturing," NBSIR 86-3312, 1986.

9.   D. Libes and E. Barkmeyer, "The Integrated Manufacturing Data Administration System, (IMDAS) - An Overview," *International Journal of Computer Integrated Manufacturing,* **1, 1,** 44-49, 1988.

10.   S. Su, "Modeling Integrated Manufacturing Data with SAM*," Computer, 34-49, January, 1986.

11.   E. Barkmeyer and J. Lo, "Experience with IMDAS in the Automated Manufacturing Research Facility," NISTIR 89-4132, 1989.

12. J. Day and H. Zimmerman, "The OSI Reference model," *Proceedings of the IEEE*, **71**, 102-107, 1988.

13. MAP and TOP Version 3.0 Specifications, Society of Manufacturing Engineers, Detroit, MI, USA, 1988.

14. S. Rybczynski, E. Barkmeyer, E. Wallace, M. Strawbridge, D. Libes and C. Young, "AMRF Network Communications," NBSIR 88-3816, 1988.

TECHNOLOGY TRANSFER FROM THE AUTOMATED MANUFACTURING RESEARCH PROGRAM AT THE NATIONAL INSTITUTE OF STANDARDS AND TECHNOLOGY

PHILIP NANZETTA

Office of Manufacturing Programs
Manufacturing Engineering Laboratory

I. Introduction

In 1988, the National Bureau of Standards (NBS) became the National Institute of Standards and Technology (NIST). In this chapter, we use the name "NBS" in references prior to the change and "NIST" for references after the change and for mixed references. This approach will give the reader a feel for the changes which the institution has undergone.

In this chapter we describe the development of the Automated Manufacturing Research Facility (AMRF) at NIST, its technology transfer mechanisms and results, and the new technology transfer programs which NIST is undertaking to develop. The organization is roughly chronological. There are some far reaching and interesting outcomes from the technology transfer designs that originated in the AMRF, including, for example, catalyzing the transformation of NBS into NIST; spawning major programs like the Navy's RAMP; development of the robot control and data architecture for NASA's space station; laying the technical foundation for a very successful line of high accuracy coordinate measuring machines; supporting the upgrading of manufacturing in naval shipyards; and helping to bring advanced manufacturing technology to America's small and medium sized firms.

This chapter also touches on the focus we see for the AMRF over the next several years, and on our thoughts for technology transfer mechanisms that we believe will contribute to strengthening the nation's manufacturing base. The early AMRF work dealt with technical problems of control, machine accuracy, data management, communications, scheduling and planning, sensors, robotics, and material handling. In the beginning, we adopted conventional methods for moving our results out of the laboratory and into use on the shop floor. Now it is clear that we need to devote a part of our resources to developing an understanding of *how to* move technological results into commercial practice quickly and surely. We are fortunate that the shapers of our institution have seen this need so clearly in defining the new missions for NIST.

II. The National Bureau of Standards

"If men are to accomplish together anything useful whatever they must, above all, be able to understand one another. That is the basic reason for a National Bureau of Standards."
— Vannevar Bush in "Measures for Progress" [2]

The National Bureau of Standards was begun in 1901 to provide a foundation for American commerce through the science of measurement and the practice of standards. It quickly became a preeminent research laboratory and a source of reference standards widely used by American industry. Today it is the only U.S. government laboratory with the sole mission of supporting the nation's commerce.

Because of its reservoir of expertise in all the physical sciences, NIST has been called upon over the years to investigate major disasters and technical issues of national interest, and to provide expert recommendations. Since 1971, NIST scientists investigated ten earthquakes, including ones in Mexico City, Soviet Armenia, and California. In cooperation with the National Science Foundation, the Federal Highway Administration, and the California Department of Transportation, NIST tested bridge columns to help develop improved criteria for design to better resist earthquake forces. NIST has a long history of investigating structural failures, focusing on determining the most probable technical causes of the failures. Such recent investigations include the Ashland Oil tank failure, construction failure of the L'Ambiance Plaza apartment building in Bridgeport, Conn., collapse of walkways in the Hyatt Regency hotel in Kansas City, construction failure of a reinforced concrete cooling tower at Willow Island, WV, and explosion of a chemical tank at the Union Oil Refinery near Chicago. NBS was directed by Congress to assess the possibility of structural deficiencies in a new U.S. embassy building in Moscow. In a different vein, NBS scientists studied a means proposed by CBS Records to prevent copying of the new digital audio tape. The conclusion of the study was that the system does not achieve its stated purpose in terms of protection and degradation of tape sound. In none of these activities does NIST have any regulatory role. It is called upon for neutral technology expertise.

NBS has *done* technology transfer from the beginning, long before the phrase achieved popularity. It has used a variety of means that basically fall into two categories, *distribution* and *networking*. Many of the NIST mechanisms for service consist of distributing the results of internal expertise to an appropriate body of users. In this category we place sale of standard reference materials, provision of calibration services, and publication of scholarly works.

Under the heading of networking, NIST engages in technology transfer through interactions between its scientists, engineers, and technicians and those of companies that wish to make use of the technology or contribute to the technology. Networking activities include participation in standards committees and scientific conferences; invitation of research associates, guest workers, postdoctoral fellows, and others to work at NIST side-by-side with NIST personnel on projects of common interest; exhibition of technology through open house and on-the-road displays of laboratory results; and many other ways.

The special strength of NIST in technology transfer derives from three factors: its indisputable expertise in the technical areas upon which it concentrates, its nine decades of experience with networking in an open environment, and its rich history of reliable contribution in times of national need.

III. Plan for a National Automated Manufacturing Laboratory

In 1980, the management of what is now the Manufacturing Engineering Laboratory prepared a detailed plan, "Proposal for a Productivity Enhancement Program for the Discrete Part Industries", for a National Automated Manufacturing Laboratory. This plan, which would have required $100 million per year for full implementation, was not funded as a whole. Nevertheless, the directions of the NIST automation program, for both research and technology transfer, were set by this plan. Research thrusts in machine tool structures, robotic assembly, design for automation, optimization, computer integrated manufacturing, sensor systems, system robustness, and quality in the automated environment, all set out in the plan, have been pursued at NIST. Technology transfer mechanisms including a productivity extension service, "automation familiarization centers", and an office of productivity related inventions are reflected in the present State Technology Extension Program, Manufacturing Technology Centers Program, Energy Related Inventions Program, and Office of Technology Commercialization.

The Plan identified foreign competition as a serious threat and proposed that attention be focused in three areas:

"*1) A well planned program to increase understanding and exploration of existing technology now and in the future.*

"*2) A long-term research and development effort for continued innovation in manufacturing technology.*

"*3) A strong program to raise the general level of education in the manufacturing sector to both reduce the perceived risk of new technologies and expand understanding of existing technologies.*"

These objectives, **technology transfer**, **research**, and **education** parallel the conclusions from a GAO study, "Manufacturing Technology - A Changing Challenge to Productivity" [9], also cited in the Plan, which concluded that

"*Significant short-term benefits are possible through improved diffusion of the available technology. For long-term sustained productivity increases, research and development is necessary to find new methods, and to refine existing technology so that it can be economically used outside the few highly capitalized, high technology firms.*"

IV. The Automated Manufacturing Research Facility

The Automated Manufacturing Research Facility (AMRF), conceived along lines described in the Plan for a National Automated Manufacturing Laboratory, was started in 1982 as a joint project with the Navy Manufacturing Technology program. The objective on the part of NBS was to build a research testbed and a reservoir of expertise in automated manufacturing which would provide the foundation for a major automation program. The Navy's aim was to accelerate the introduction of automated manufacturing into its vendors' and its own manufacturing and rework facilities. The physical layout and research programs of the AMRF are described in another chapter of this book.

The AMRF and the automation program that it supports is a highly visible national program in automated manufacturing. From the beginning, it has been widely supported by important sponsors and it has enjoyed technical contributions from many co-workers outside NIST. There are no success stories from the AMRF which are untouched by this participation. In fact, we would judge one of the major successes of the program, a success which flows naturally from the networking environment of NIST, to be the extent to which the AMRF has served as a focus for interaction among the many participants. We do not forget, and the reader must not forget, that when we attribute impacts to the AMRF, the whole team, external as well as internal, has made the impacts happen and deserves the credit.

Foremost among the sponsors of the AMRF is the Navy Manufacturing Technology Program. Jack McInnis, director during the founding of the program, and Steve Linder, the current director, provide crucial funding without which the AMRF would not exist. They also provide access to Navy applications which help the program bring its technical resources to bear on valuable applications, and which serve to focus the research on important matters.

Results from the AMRF have influenced the directions of manufacturing by Navy suppliers such as Texas Instruments and General Dynamics, who have participated as Research Associates in the NIST programs. Projects have been undertaken by the AMRF in the Naval Shipyards, Naval Air Depots, and other internal Navy facilities. The Air Force, Army, DARPA, NASA, SDIO, DOE, Interior Department, and other government agencies have also participated in supporting work in the automation program directed toward their needs. Research topics include autonomous vehicles, robot control architectures, full factory automation architectures, high precision machining, machining of exotic materials, and ultra precision angle measurement. We mention this work here because of the very important and very large effect of cross-fertilization among projects and the means by which NIST projects achieve critical mass through multiple applications of fundamental theoretical findings.

V. Transfer of Specific Items of Technology vs Models

Technology transfer is usually discussed in terms of well-defined, clear-to-document specific instances such as accomplishment of a particular project with a known cost and carefully projected return on investment. Licensing of patents, cooperative research and development agreements, assistance with implementation of new equipment on a shop floor, all of these are specific and straightforward to document. The AMRF has accomplished a great deal of this type of specific transfer. The tables on the following pages display some examples.

INDUSTRIAL RESEARCH ASSOCIATES OF THE AUTOMATED
MANUFACTURING RESEARCH PROGRAM

Acme-Cleveland Corporation
Allen-Bradley Company
Arthur Andersen & Co.
Bendix Corporation, Kansas City Div.
Boeing
Brown and Sharpe Manufacturing
Case Consulting
Cincinnati Milacron, Inc.
CMX Co.
DACOM
E. Fjeld Company, Inc.
EDAX International, Inc.
Electronic Measuring Devices, Inc.
Factrol, Inc.
FMC Corp./Northern Ordinance Div.
GCA Corporation
GCA Corporation/Tropel Div.
General Dynamics, Pomona Div.
General Electric Co.
Hardinge Brothers, Inc.
Hewlett Packard
Honeywell, Inc.
Hurco

IBM
Kennametal, Inc.
Lockheed Missiles and Space Co.
Management Collaborative Group
Mare Island Naval Shipyard
Martin Marietta Corp.
Meridian Corp.
Metcut Research Associates, Inc.
Monarch Machine Tool Co.
NASA
National Tooling and Machining Assoc.
Ontologic, Inc.
Qint Database Systems Corp.
Science Applications, Inc.
Structural Dynamics
Tandem Computers, Inc.
Texas Instruments, Inc.
Transitions Research Corporation
Vickers Instruments
VLSI Standards, Inc.
Warner and Swasey Co., Sheffield Div.
Westinghouse
White-Sundstrand

PATENTS FROM AUTOMATED MANUFACTURING RESEARCH PROGRAM

NAME	AUTHOR(S)
"Split-Rail" Parallel Gripper	Vranish/Bunch/Johns
Inclined Contact Recirculating Roller Bearing	Slocum
Multi-Port Hydraulic Commutator Valve	Slocum/Peris
"Quick Change" System for Robots	Vranish
Robot End Effector	Slocum/Jurgens
Pipelined Image Processors	Kent
Magnetorestitive Skin for Robots	Vranish/Schwee/Goetz
Rotating Tool Wear Monitoring Apparatus	Yee/Blomquist
Interferometric Angle Measuring Device	Lau
Automatic Laser Interferometer Tracking Device	Lau/Hocken
Device of In-Line Measurement of Lathe Cutting Tools	Peris
Method and Mechanism for Fixturing Objects	Slocum/Peris
Method and Mechanism for Inserting Parts	Slocum/Peris
Local Buffer Storage for Robotic Workstation	Reisenauer/Gaver/Slocum
Ultra-Sonic Device to Monitor Surface Roughness and/or Tool Wear	Blessing/Eitzen
Automated Collect Changer	Reisenaur/Lee/Jurgens
Multiple Actuator Hydraulic System	Slocum/Peris
Rotary Control Valve	

PRODUCTS FROM AUTOMATED MANUFACTURING RESEARCH PROGRAM

NAME	COMPANY	DESCRIPTION
CMM Accuracy	Sheffield	Software Error Correction
Drill-Up	Valeron	Drill Wear Detector
Drill-Up	Technovations	Drill Wear Detector
Validator CMM	Brown & Sharp	Software Correction
Apollo CMM	Sheffield	Software Correction
KV Quick Change Tools	Kennametal	Tooling System
Quick Change	Medical Robotics	QC Wrist
Quick Change	Milacron	QC Wrist
Gripper	Lord Corp.	Robot Gripper
Error Correction	Hardinge	Lathe Error Correction
Error Correction	Brown & Sharp	Mill Error Correction
Tool Setting	Hardinge	On Machine Tool Setting
Robot Vision	Automatix	Structured Light
PIPE	ASPEX	Image Processor

SUBSYSTEMS FROM AUTOMATED MANUFACTURING RESEARCH PROGRAM

NAME	LOCATION
Servo Vise	Goodyear Tire
Emulator (HCSE)	General Electric, BB & N
Controllers	Boeing/Computer Services,FMC, Timken
Process Planner	Bendix Kansas City, Texas Instruments
Facility Software	South Carolina Research Authority
Robot Vision	Northrup
IMDAS	Lehigh
Robot Vision	Digital Signal
Robot Control System	Honeywell, Martin Marietta, Allen Bradley, Rensselaer Polytechnic Institute
CAD Directed Inspection	Rensselaer Polytechnic Institute
Feature Driven Machinery	Rensselaer Polytechnic Institute

These specific instances are important, of course, for the organizations involved with them, and they are important as concrete illustrations of the power of the technology being explored. Their aggregate dollar impact is even quite large. But their effect at the national level, standing alone, is not far-reaching. By themselves, these specific projects hardly suffice to justify the expenditure of tax-payer's money on a major program like the AMRF. The real, high-value payoffs come in the form of models that the program provides for other programs, for entire technology areas, and for major implementations. Unfortunately (for the sake of budget justifications) the impact of these models is difficult to trace specifically and the lineage of the impacts is hard to attribute to its root causes.

The successful implementation of new technology does not follow from a single cause. The most clever new idea has no value unless it is adopted and used. That will only happen if someone takes the risk of putting capital behind it, which will take place only if the technology is introduced in a way that stimulates confidence; it must be tested or tried in other applications. Finally, the market environment must accept the new technology. Because of all these parents of success, it is not possible to point to "the" reason for a major technological advance. In splitting up the credit, it is not possible to decide reasonably on the share to attribute to each parent.

In looking at the outputs of a program like the AMRF, the really important impacts are of this type. Fortunately, there are a few cases in which the record is clear enough to help illustrate this point.

VI. Technology Transfer Through Models for Programs

Modern practitioners say that technology transfer is a contact sport. This may be true in many cases at the level of the ultimate consumer. But there are enough examples to suggest that technology is also transferred by providing models which others can use in focusing their thinking. Such models provide a framework to use in analyzing a new situation and in selling it to potential sponsors. Observing the success of one example also serves to reduce the concerns for risk and may motivate further steps, given that a familiar foundation has been demonstrated. The examples which follow are programs that build on the model demonstrated by the AMRF.

Since the demonstration of the completed AMRF shop floor facility, there has been a flowering of commercial implementations of automation software. Automated manufacturing research, teaching, and demonstration centers, embodying the principles developed in the AMRF, have been established at universities and technical colleges. A Navy facility known as RAMP, designed to insert advanced manufacturing and logistics principles into Navy procurement of spare and repair parts, is our first example.

A. The Founding of RAMP

The Naval Supply Systems Command was one of the initial sponsors and technical partners of the AMRF. The research program explored the idea of eliminating much of the Navy's vast stockpiles of spare parts by instead organizing the data to make those parts in computer-readable form and constructing facilities which could make the parts from the data. In shorthand, "warehouse data, not parts". The AMRF was the original model for the manufacturing facility and IGES, the Initial Graphics Exchange Specification developed at NBS, was the original data description standard.

This is an application with tremendous leverage. The value of the Naval stockpile of parts is huge and on the average, only a very small percentage of these parts will ever be used. Yet, they must all be available because the need is not predictable. Under the "warehouse data, not parts" paradigm, only those parts which are needed must actually be made, yet the Navy can still obtain needed parts.

As described in internal memoranda of July, 1984, the NBS Parts on Demand (POD) system was to be "an integrated system for computer assisted entry, transformation, verification, archiving, transmission, and revision of geometric and non-geometric part design information, process plans, NC programs, or instructions for required unit operations, and inspection plans; and for minimal production management". Data formats for part description played a key role in early discussions of POD.

By early 1985, the name was changed from POD (an acronym which, unfortunately, was already taken by Printing on Demand) to Rapid Acquisition of Manufactured Parts (RAMP). The RAMP objective, presented by Navy planners was twofold:

"Integration of advanced flexible manufacturing technology into the Navy logistics system

"improved availability of spare parts to meet fleet readiness and sustainability objectives."

The Navy planned to demonstrate a fully operational parts on demand system based on NBS development of manufacturing and interface standards and commercial development of administrative support computer systems.

The initial views of a RAMP shop floor facility, in fiscal year 1987, were based on the completed workstations of the AMRF, including horizontal machining station, vertical machining station, turning station, and quality assurance station. It included an emulator and moved materials via an A.G.V. This resemblance is more important as an illustration of the strength of *models* than as a reflection of deep planning. Further thought by the designers of RAMP convinced them to use a different floor plan, more suitable for their objectives.

Along with the model for a shop floor system came NBS influence on data standards. A 1987 statement of work entitled "Define PDES 1.0 as integral to CALS Phase II, Part 2" makes the connection clear, as follows:

"The Product Data Exchange Specification (PDES) is designed to permit the communication of product definition information directly between heterogeneous computer systems. Development of this standard has been jointly undertaken by the National Bureau of Standards and (on a voluntary basis) industry and government."

Throughout the development of the RAMP program, NBS advisers strongly urged the Navy and RAMP to base their system on the emerging PDES standard. This advice was taken, and now PDES development is a sister program of RAMP within the South Carolina Research Authority (SCRA). PDES underlies all of the work in RAMP, and NIST, with its National PDES Testbed and CALS program, plays a companion role with SCRA in development and testing of PDES.

An interesting sidelight: It was at the 1987 ground breaking ceremony for the RAMP facility in South Carolina that the new name National Institute of Standards and Technology and the augmented mission were first publicly announced by Senator Ernest F. Hollings.

B. Navy Centers of Excellence

The AMRF was the first Navy Center of Excellence. As the AMRF was being developed at NBS, the Navy Manufacturing Technology Program studied the impacts of various approaches to its goals of reducing costs, improving turn-around time, and preserving the nation's independence from foreign suppliers of critical components. Based on that analysis, it decided to focus its resources increasingly into Centers of Excellence. The Navy has now established three additional such centers, in electronics, composites, and metalworking. These centers are, respectively, the Electronics Manufacturing Productivity Facility (EMPF) located in Indianapolis, Indiana; the National Center for Excellence in Metalworking Technology (NCEMT) located in Johnstown, Pennsylvania; and the Center of Excellence for Composites Manufacturing Technology (CECMT) located in Kenosha, Wisconsin.

VII. Technology Transfer by Models of Applications

Whether the company is Boeing or General Electric, or a small manufacturing firm, or a major defense contractor, when the time comes to consider introduction of new technology, the important technical and economic issues must be addressed by individual people. The work of these people is accelerated by having reliable models for the applications they are considering. Since the initiation of the AMRF, tens of thousands of people have visited its laboratories and explored with its scientists and engineers the issues with which they were concerned back home in their companies. The following examples are quoted from "The National Bureau of Standards' Automated Manufacturing Research Facility (AMRF)" by William P. Meade [5].

"*Mr. Gene Foster, Computer Integration manager, Manufacturing R&D, Boeing Commercial Airplane Company, The Boeing Company, Seattle, Washington, noted:*

'*I guess the greatest value of the AMRF is its pioneering approach to the overall system control hierarchy. Our control architecture is patterned after theirs as well as that of ICAM which has also adopted the AMRF hierarchical approach.*'" *([5] page 29).*

"*Mr. Bob Solberg, Chairman, Factory Automation Council, Boeing Computer Services, the Boeing Company, Seattle, Washington, stated:*

'*The AMRF provides a good architecture for factory automation which we are using in our internal factory automation program. We are picking up a lot of concepts such as feedback loops, world modeling and planning.*

'*We are developing a generic cell based on the AMRF architecture.*'" *([5] page 33).*

"*Mr. Robert D. French, Manager T700/CT7 Engine Programs, General Electric Company, Aircraft Engine Business Group, Lynn Massachusetts, noted:*

'*In 1984 I joined the Aircraft Engine Group of General Electric as the Manager of Automation for a flexible turning center to be built in Lynn, MA, becoming involved at the time when the concept was to be turned into reality. The NBS/AMRF hierarchical control structure was one of the models used in designing the computerized control system for the plant. More than this, the AMRF provided us with a reference, something we could check ourselves against. I also had key people spend a day at the AMRF to recalibrate themselves in specific technology areas. Finally, some of our equipment suppliers, such as Monarch, were supportive of the AMRF and understood where we were heading with the new Lynn facility.*'

'*Some of the things being done at the AMRF are profound in my view. While the overall thrust of the AMRF has been widely publicized it tends to overshadow details. For example, AMRF researchers introduced error maps to facilitate the use of computer-based temperature correction to upgrade the precision of a milling machine.*'" *([5] page 36, 37).*

"Mr. Chad Frost, President, Frost Incorporated, Grand Rapids, Michigan, stated:
'The AMRF showed us what was doable. They did not show us how to do it. Rather, our thinking was confirmed. We obtained assurance that our approach was feasible, thus allowing us to launch a $4.5 million project we could afford rather than the $12, $15, and $30 million projects proposed by some vendors. During implementation, the AMRF did provide some hand holding, providing assurance that certain things could be done, such as tie together two dissimilar pieces of equipment.'" ([5] page 43).*

"A spokesperson for a defense-related firm related:
'I have the responsibility for implementing a new CIM project. We were just getting into the project when we visited the AMRF. It was our first real-world exposure to automation. What we saw at the AMRF blew our socks off! The AMRF reinforced many of our concepts and exposed us to what could be done.*

'Since our visit to the AMRF we have moved forward with the implementation of our CIM project. We have visited other plants and met with many vendors. While our approach to CIM may be different than the AMRF with respect to hardware, networks and redundant systems, the systems architecture, the backbone of the project, is a direct pickup of the systems architecture we saw at the AMRF.'" ([5] page 43).

VIII. Technology Transfer by Models for New Technology Families

Above we described several specific items of technology which have been developed in the AMRF, patented, and then licensed for implementation by commercial concerns. In contrast, the examples presented in this section show a broader picture, illustrating the transfer by models for new technology families. The first is *software error correction* and the second is the *hierarchical control architecture*.

NBS research on software error correction began in the late 1960s with work on coordinate measuring machines (CMMs). Observing that good CMMs had a much higher degree of *repeatability* than they had *accuracy*, the NBS team first fitted a computer controller to a manually controlled CMM. They then mapped the errors of the machine (this work alone deserves a whole book of description) and arranged the controlling programs to make reference to this error map so as to factor the errors out of the machine. In this way, it was possible to improve the accuracy of the machine by more than an order of magnitude. [4] contains a more detailed description. In this case, the errors that were mapped were the inherent kinematic errors at constant temperature. Subsequent work has been carried out on temperature-based errors [3]. The work has been extended from CMMs to turning centers and machining centers. As the theory was being developed, NBS metrologists used the results in high-accuracy production machining for the U.S. Bureau of Engraving and Printing. The principle of software error correction is now applied very widely, not only to machine tools but to such diverse applications as vision-based metrology, in which inherent errors of a video camera are mapped and corrected in software.

[5] contains the following description of a complete line of modern CMMs which are based on the principles of software error correction:

"Mr. John A. Bosch, Vice President & General Manager, Sheffield Measurement Division, The Warner & Swasey Company, A Cross & Trecker Company, Dayton, Ohio, noted "software accuracy enhancement is a milestone in machine accuracy:

'Our recently introduced Apollo coordinate measuring machine (CMM) was technology transfer in action. It includes an enhanced version of software error correction developed at the AMRF by Bob Hocken and his associates.

'The Apollo includes microprocessor-based accuracy enhancement through software error correction developed at the AMRF. The error correction through software is revolutionary and its genesis came from the National Bureau of Standards' Automated Manufacturing Research Facility. Certainly the methodology we used to realize the commercial application of software correction was different. That is to be expected, the Bureau and its AMRF researchers are not attempting to develop market-ready products.'

. . . 'Within Sheffield we have made a cost analysis of the Apollo versus traditional forms of manufacture and assembly. I will not tell you the savings we realized but will tell you it, software correction, is the most significant thing that has evolved in the use of CMMs. We are now providing highly dynamic machines instead of static structures.'

The development of the hierarchical control architecture for shop floor systems and robotics began at NBS with research on the architecture for early robot controllers. Under this approach, a complex task is decomposed into simpler tasks and grouped into levels, with higher levels managing broader, less detailed aspects of the task and lower levels accepting commands from above and handling successively finer detail. The concepts were tested for the AMRF by development of a Hierarchical Control System Emulator, and were then implemented as the fundamental architecture of the AMRF. As described in examples above, this architecture serves as the basis for production factory control systems. It has been adopted by NASA [1], under the name of NASREM, as the architecture for the telerobotic control system for the space station. It serves as the underlying architecture for a standard, SARTICS, which is being developed for control of intelligent machines. And it is the basis for design of the important Air Force Next Generation Controller.

In both of these examples, the research and fundamental genius of invention were motivated by solution of very specific problems. The work was carried out by scientists who were deeply immersed in their professional fields. It was only with years of development, testing, and application that the full potential of these ideas began to become clear. The AMRF provided a rich environment of people, equipment, observers, and production applications in which the basic ideas were tried out and expanded. The position and background of NIST enabled the transfers.

IX. Technology Transfer by Models for Standards

The AMRF does not establish any standards, but it does provide technical input in the form of research results and testing of concepts to support voluntary industry standards-setting. AMRF personnel sit on the relevant voluntary committees and in some cases, NIST serves as the secretariat. AMRF contributions to the standards process represents one of the most significant modes of technology transfer, with high leverage in terms of dollar impact and international competitiveness. The accompanying tables present some of the standards-based involvement of AMRF personnel.

STANDARDS FROM AUTOMATED MANUFACTURING
RESEARCH PROGRAM

IN PLACE

Initial Graphics Exchange Specification ver. 1.4	(ANSI Y 14.2 M)
Characterization of CMM	(ANSI B89.1.12)
Surface Texture	(ANSI B46.1 - 1985)
Automated Interchange of Technical Information	DoD
Digital Representation for Communication of Product Data Application Subsets	DoD
Markup Requirements and Generic Style Specification for Electronic Printed Output and Exchange of Text	DoD
Raster Graphics Representation in Binary Formatr	NCSL
Digital Representation for Communication of Illustration Data	NCSL
CGM Application Profile	CALS, DoD

STANDARDS FROM AUTOMATED MANUFACTURING
RESEARCH PROGRAM

IN THE MILL

Models for Factory Architecture	ISO TC 184 SC 5
Industrial Automation	ANSI Panel (IAPP) (Chair)
Information & Communication, Robots	RIA R15.04 (Chair)
Robotics and Automation	IEEE R&A Council (Chair)
Robot Performance	RIA R15.05
Data Exchange Standards	ANSI X3
Manufacturing Automation Protocol	EIA
Performance of Machining Centers	ANSI B5 TC52
Standard for the Exchange of Product Model Data	ISO TC 184 SC4
Remote Data Access	TC X3 H2.1
Interchange of Large Format Tiled Raster Documents	NCSL (Chair)

X. Omnibus Trade and Competitiveness Act of 1988

In 1988, NBS was transformed into the National Institute of Standards and Technology (NIST). Its mission was augmented by the addition of a set of extramural programs organized explicitly to transfer NIST and federal laboratory technology to U.S. based manufacturers, especially to those who were out of reach of traditional means of transfer. Yet another impact of the AMRF is its influence on the course of institutions. According to many observers, the automation program at NBS was a significant factor in the transformation of NBS to NIST. To quote from a report by the National Academy of Sciences [8]:

"The AMRF has for years served as a platform to develop needed technology for flexible, integrated, and automated manufacturing of discrete parts. Originally conceived as a testbed for integration of advanced automation, it successfully proved a number of concepts and eliminated less promising ones. It has played a significant role in the identification and development of emerging technologies in manufacturing. It has had considerable influence on various private efforts throughout the nation. It was also the catalyst in the legislative process that resulted in the Technology Competitiveness Act" (the section of the Omnibus Trade Act of 1988 which reauthorized NBS and NIST). ([8], page 232)

The Trade Act added a new method of technology transfer to the two existing methods of distribution and networking. This new method extends the power of networking by calling upon NIST to establish and fund external organizations to serve as transfer agents, thus providing a new order of magnitude to the *leveraging* of technology transfer.

The new programs include the Advanced Technology Program (ATP), a State Technology Extension Program, and the Manufacturing Technology Centers (MTC).

The ATP program provides funding on a competitive basis to industry to accelerate the development of pre-competitive, generic technologies with high commercial potential. This is a valuable channel to the commercialization of AMRF-originated technology. One of the ATP awards in the first round of competition went to a company, Saginaw Machine Systems, for the further development of machine tool accuracy enhancement techniques originated at NIST and other laboratories. The focus is on compensating for thermal effects in the machine tool.

The STEP program assists the states in promoting the use of improved technology by their industry. Several of the initial awards involved linkages with Manufacturing Technology Centers.

The Manufacturing Technology Centers Program was established to transfer advanced manufacturing technology to small and medium sized U.S. based manufacturers, focusing especially on technology that is available from NIST and other federal laboratories. At the time of this writing, there are five MTCs, as described below.

The Great Lakes Manufacturing Technology Center (GLMTC) is hosted by the Cleveland Advanced Manufacturing Program (CAMP), located on the campus of Cuyahoga Community College in Cleveland, Ohio. It receives substantial financial support from Ohio's Edison Program. GLMTC is affiliated with Case Western Reserve University, Cleveland State University, Cuyahoga Community College, and Lorain Community College. It supports the Advanced Manufacturing Center at Cleveland State University as its principal source for developmental work. In 1990, GLMTC reported contacts with 2,855 new companies; 312 plant visits; and 26,300 participants in trade shows, public forums, and professional organizational meetings. It completed 50 major technology transfer projects with an estimated value of $80.8 million.

The Northeast Manufacturing Technology Center (NEMTC) is hosted by Rensselaer Polytechnic Institute's Center for Manufacturing Productivity in Troy, New York. It receives some support from the state of New York, in-kind assistance from the New York State Industrial Technology Extension Service, New York State Science and Technology Foundation, and fees from clients. NEMTC is affiliated with the Hudson Valley Community College for outreach, and with organizations in the states of Maine (Maine Production Technology Center) and Massachusetts (TECnet and Machine Action Project) as "satellite" operations. Its technology emphasis is on Computer-Aided Design, Engineering, and Manufacturing, Supplier Development Programs, ISO9000 quality certification, and cooperative development programs using NIST technology.

The Southeast Manufacturing Technology Center (SMTC) is hosted by the University of South Carolina together with the South Carolina Technical College System. It receives substantial support from the state of South Carolina, and uses the technical colleges to reach the approximately 2,000 companies in the state with direct services. The technology emphasis of SMTC is solids modeling, finite element analysis, stereolithography (for rapid prototyping), electronic data exchange, computer networking and networked delivery of services. It has developed training packages in specialty areas for upgrading technical college faculty. It is working to develop the Southeast Manufacturing Network to link first South Carolina and then the southeast region for delivery of courses, processing of technical queries, and support for business development.

Two new MTCs have been added recently. The Midwest Manufacturing Technology Center, hosted by the Industrial Technology Institute in Ann Arbor, Michigan, and the Mid-America Manufacturing Technology Center, hosted by the Kansas Technology Enterprise Corporation, were recently selected.

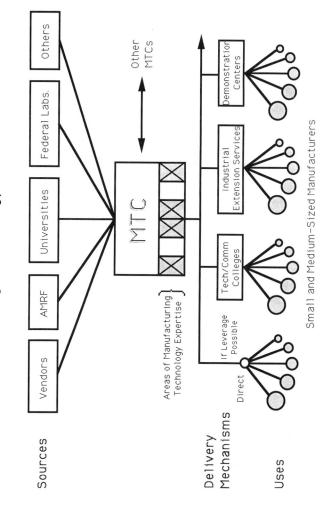

Manufacturing Technology Center Model

Sources

| Vendors | AMRF | Universities | Federal Labs. | Others |

Other MTCs

MTC

Areas of Manufacturing Technology Expertise

Delivery Mechanisms

If Leverage Possible

Direct

| Tech/Comm Colleges | Industrial Extension Services | Demonstration Centers |

Uses

Small and Medium-Sized Manufacturers

Figure 1. The Manufacturing Technology Center model. The center is the focal point for technology transfer between sources and users of technology.

The MTC program follows a model of technology input, reservoir, and delivery shown in the diagram on the preceeding page.

It is interesting to explore for a moment the original vision for the MTCs and to highlight several ways in which this technology transfer vehicle has gone beyond the original expectations by catalyzing new relationships between natively distant organizations. Prior to the Trade Act, the AMRF could transfer technology effectively to large manufacturers, who could afford the resources and risk of moving results from the laboratory to the production floor. The AMRF could also transfer technology to smaller firms which were vendors of the technology; their business was to transform partially developed technology into a product. The AMRF could do little to transfer technology to the smaller firm which was just a user of the technology. They did not have the resources and could not afford the risk of bringing unproven technology to their production floor. The MTC program was designed to serve this very important sector of American manufacturing.

Each of the MTCs has programs and personnel whose sole mission is the transfer of advanced manufacturing technology to smaller manufacturers. They each have formed rich networks of technology sources and service providers. Their networks generally include technical or community colleges, state technology extension services, state economic development offices, research universities, large companies which want to strengthen their supplier base, and other nonprofit organizations which share the mission of technology transfer. In the process, cooperation has been sparked between research universities and community colleges, between state economic development organizations and small in-state companies. This impact reaches into the classroom and onto the production floor. The MTCs are a driving force for enrichment of communications among state level entities.

XI. Internally Sponsored Programs

As part of its internal efforts to extend the reach of technology transfer, NIST has undertaken to develop several internally sponsored programs. The most visible of these to date is the Shop of the 90s, which is described in a separate little chapter in this volume. This program has evolved hand-in-hand with a modernization effort for NIST's internal instrument shop. The objective of the Shop of the 90s is to test, demonstrate, and encourage adoption of commercially available, off-the-shelf computer based technology for small shops. A demonstration facility has been assembled at NIST in cooperation with industrial partners who supply computer hardware and software suitable for small shops. The program has been led by the top management of the NIST instrument shop, and has provided a number of presentations designed to stimulate small firms to consider practical technology upgrades in their own operations.

The Opportunities for Innovation project is aimed at the smaller technology-based firm. It is issuing a series of NIST monographs designed for the technical staff of such a firm, providing it with an industry-wide perspective on the best opportunities for new business endeavors in a technology-driven marketplace. The monographs deal with both mature technologies and emerging technologies, covering applications in automotive, aerospace, and consumer goods markets. A workshop program will be matched to each monograph to provide a richer interaction between the technical experts engaged for the program and the end users.

XII. What it takes to transfer technology

As the examples described above make clear, there are many dimensions and layers to the transfer of technology. Sometimes, it is sufficient for an end-user simply to be made aware that a particular approach is possible. At other times, a well-developed and well-documented technology exists as a discrete entity and must be licensed, or a product embodying the technology is available in the marketplace and must be procured. Some adoptions of technology involve the manufacturing processes and others focus principally on the product.

For purposes of understanding technology transfer from a facility such as the AMRF, we must consider separately the three types of company discussed above, large companies, small technology vendors, and small technology users.

Large companies have the resources, if they decide to utilize them to acquire new technology, to draw upon university and government research laboratories to obtain technology in its early stages of development. The Research Associate program at NIST has been very successful over the years in supporting this form of technology transfer. Personnel from the company work at NIST for a period of time on a project which is of interest to them and to NIST. The scientific results are published in the open literature, valuable intellectual property is protected by patents according to the source of inventions, and the research associate returns to his or her company with a deep understanding of all aspects of the technology. Recent legislative changes are strengthening this tool for utilizing government technology sources to benefit American industry.

Large companies can utilize the research associate approach because they can afford to lose the services of one of their best people while they are out associating, and because they can devote resources later to carry a technological idea through the developmental stages to practical application. They can afford some risks in implementing new technology and testing new products.

Small companies which are vendors of technology have also been able to work closely with AMRF personnel on development and transfer of research results. When they engage in a research associate program that is properly focused, the NIST laboratories serve as an extension of their own. Their key personnel may well be stationed at NIST for a period of time. Their business is to carry out the commercialization process.

Small companies which are users of the new technology can not, in general, afford the cost, disruption, or risk of starting at the laboratory level. They need, at the very least, to be able to procure the components already developed, tested, and fully supported. It remains for them to understand the available technology, select and integrate the components, and exploit the technology throughout their operations. For large companies, this function may be performed by systems integrators. In our experience, small companies do not represent an attractive market for systems integrators because the front-end costs of reaching them are too high in relation to the price of the system they can afford.

The high level of autonomy inherent in the American culture leads to creativity and invention. But it also erects barriers that weaken national competitiveness. This cultural element heightens the need for approaches which enable cooperation and linkages between people, departments, and companies. The tools that promise to support such cooperation will also bring to aggregations of small technology-using companies some of the advantages enjoyed by large companies, pooling resources to support the costs of technology adoption.

XIII. Reaching for Solutions: Navy Applications

The Navy Manufacturing Technology Program co-sponsors the AMRF, so a significant part of the AMRF technology transfer involves implementations in Navy facilities or work with Navy vendors. Not only is the result beneficial for the Navy facility, but it also provides real tests for the ideas of AMRF engineers and scientists. It helps broaden our field of endeavor from the abstract and untested into the practical arena of applications.

After the AMRF demonstrated the success of its shop floor automation, the Naval Sea Systems Command invited NIST to build an integrated system to install in the Mare Island Naval Shipyard to make a family of machined components for nuclear submarines. This family includes 40 different pipe connectors which serve to inhibit the transmission of sound as part of the critical sound silence program for nuclear submarines. The workstation is equipped with an NC lathe that uses live tooling, a robot, an automated storage and retrieval system, and an AMRF-type control system. Current manual techniques require 17 hours to produce a RISIC part. The new workstation can manufacture the same part in as little as 20 minutes, can operate unattended, and is capable of filling part orders within 24 hours. The development of the workstation was a joint project of NIST, Warner & Swasey Company, and Westinghouse Electric Corp. The system was certified in late 1990 by test production which machined parts that were subsequently inspected and passed by the shipyard's inspection department. AMRF workers learned a great deal about technology transfer (as well as technology) from this project, which one can not count as an unqualified success. There was substantial resistance within the shipyard to using the new workstation and both shipyard and AMRF personnel had unrealistic expectations about operation of low-volume automation in unattended mode. This new knowledge is being put into practice in a second workstation project, one for Portsmouth Naval Shipyard.

The Portsmouth shipyard has a requirement for class 1 fasteners (bolts, studs, cap screws) made of k-monel, a very tough and difficult-to-machine material. The fasteners are used on board submarines in applications for which a broken fastener would threaten the survival of the vessel. These fasteners are expensive to machine and difficult to procure from outside sources. There are strict audit trail requirements, specifications for certified raw materials, and frequent inspections throughout the manufacturing process. AMRF engineers, working closely with Portsmouth personnel (note!), designed a workstation to make and inspect fasteners. Portsmouth personnel helped select the equipment, and have been resident at NIST throughout integration, testing, and programming of the workstation. When it is ready to ship, most of the Portsmouth people who will use the workstation will have worked for an interval at NIST on its development. They look forward to having it on site.

At the request of shipyard commanders at Pearl Harbor, Charleston, Puget Sound, and other Naval shipyards, AMRF personnel conducted surveys to help increase efficiency and reduce the cost of submarine overhauls by applying the results of research performed in the AMRF. The shipyard commanders estimate that it costs each shipyard at least $200,000 for each day that a nuclear submarine is in dry dock. Moreover, the longer that submarines must rest in shipyards, the more submarines the fleet must contain; the capital costs related to time in the yard are even greater than $200,000 a day. NIST engineers have suggested restructuring of some shops, addition of distributed computer control, tracking and planning, and a general upgrade of the facilities. They have assisted in equipment selection in major capital purchase programs. By the accounts of shipyard officials, these surveys have had a significant impact on operations and planning.

XIV. Some Future Directions

Throughout the 1980s, there was a need for developing, testing, and sharing many of the basic ideas of large-scale manufacturing automation. The AMRF was one of the major contributors to this effort. In the process, the AMRF developed an early experience base in technology transfer, and helped transform NBS to NIST with its new set of extramural responsibilities.

Many of the research issues of 1980 have been resolved into commercial products. Many of the demonstration requirements are served by facilities distributed across the country in universities, consulting firms, community colleges, and economic development organizations. Standards have been defined and adopted in significant areas. New mechanisms of technology transfer have been initiated.

While this chapter concentrates on technology transfer, and the size of this task is massive, it is also the case that significant technology issues remain to be investigated at the level of fundamental research. We'll describe several as examples.

In the mass production mode where operators are present, the control loops necessary for quality management are well studied. Operators can hear tool chatter or see its impact on the workpiece; machinery can produce parts within tolerance through statistical process control charts and machine adjustments; problems can be addressed through special tooling. None of these techniques works in the fully automated environment where parts are made in small lot sizes or lots of one. The AMRF's Quality in Automation project is developing quality control mechanisms which will work in this new environment [3]. The control architecture envisions multiple control loops, beginning with software correction of the machine tool. On-machine probing checks for tool wear and material problems. Post-process inspection checks for machine tool drift or other problems. Some elements of the system are just a gleam in the eye of the scientists (how do you use off-machine probing results from one part to correct the machine when it is making a different part?). Other elements are being incorporated into commercial products right now.

Other major research issues include ultra-precision machining and measurement, machine accuracy and quality, new aspects of control and sensor integration for robotics, integration of factory control and scheduling, and product data description standards and technology.

The AMRF began as a research facility and testbed for understanding standards and interfaces in automated manufacturing, and it continues to address these issues for new standards and interface issues. The AMRF is heavily involved in technology transfer programs, both "old NBS" and "new NIST". Its new research thrusts deal with quality, shrinking dimensions, new materials, intelligent machines, the information structure of manufacturing, and the technological aspects of competitiveness. An experimental new focus is being developed to understand the process of making laboratory-level technology into commercial products. The new enabling legislation for Cooperative Research and Development Agreements makes it possible to bring together researchers and commercializers for that purpose.

References

[1] Albus, James S., McCain, Harry G., and Lumia, Ronald, "NASA/NBS Standard Reference Model for Telerobot Control System Architecture (NASREM)", NIST Technical Note 1235, 1989 Edition, U.S. Department of Commerce, National Institute of Standards and Technology, 1989.

[2] Cochrane, Rexmond C., "Measures for Progress - A History of the National Bureau of Standards", U. S. Department of Commerce, 1966, Washington (Superintendent of Documents Catalog No. C13.10:275).

[3] Donmez, M. A., "Progress Report of the Quality in Automation Project for FY90," NISTIR 4536, U. S. Department of Commerce, National Institute of Standards and Technology, 1991.

[4] Hocken, Robert J. and Nanzetta, Philip, "Dimensional Metrology at the National Bureau of Standards", The Physics Teacher, Vol 21 No 8, 1983, pages 506-513.

[5] Meade, William P., "The National Bureau of Standards' Automated Manufacturing Research Facility (AMRF) - A Manufacturing Technologies Resource Center", Management Collaborative Group, Chapel Hill, NC, 1988.

[6] Nanzetta, Philip, Weaver, Asenath, Wellington, Joan, Wood, Linda "Publications of the Center for Manufacturing Engineering Covering the Period January 1978-December 1988," NISTIR 89-4180, U. S. Department of Commerce, National Institute of Standards and Technology, 1989.

[7] "The Manufacturing Technology Centers Program - A Report to the Secretary of Commerce", Visiting Committee on Advanced Technology, The National Institute of Standards and Technology, 1990.

[8] "An Assessment of the National Institute of Standards and Technology Programs, Fiscal Year 1990," National Academy Press, Washington, DC 1991.

[9] "Manufacturing Technology - A Changing Challenge to Productivity - Report to Congress", Comptroller General of the United States, LCD-75-436, Washington, DC, 1976.

MANUFACTURING IN THE TWENTY-FIRST CENTURY: A VISION

JOHN A. SIMPSON

Manufacturing Engineering Laboratory
National Institute of
Standards and Technology

I. THE GOAL

The goal of manufacturing in the 21st century will be to produce World-Class Product for World-Class Customers.

World-Class Product: Product that has the highest "fitness for intended use" permitted by available technology at a price commensurate with its worth, and available at the time of need, or desire, of the customer for that product.

World-Class Customers: Customers who identify needs early, welcome innovation, have a strong sense of worth and who are satisfied by nothing less than World-Class Product.

CONTROL AND DYNAMIC SYSTEMS, VOL. 45

II. BACKGROUND

At no time in history since 1850 has it been as easy to predict, with reasonable assurance, what manufacturing will be for the next 20 to 50 years.

As discussed in detail in the chapter **Mechanical Measurement and Manufacturing**, by 1850 the enabling technology was all in place, in primitive form, that would support the First Manufacturing Paradigm (FMP). From the Woolwich Arsenal, in England had come the machine tool, from France the mechanical drawing, from Eli Whitney and the Colt Armory in Hartford CT, had come fixed jigs and tooling and factory organization that allowed the use of relatively unskilled labor to manufacture product as complex as a musket. The next 100 years saw the development of the FMP, culminating in the great mass-production factories of Ford, General Motors, and the munitions and armament plants of World War II.

Since 1950 we have seen the development of new enabling technologies which have given rise to a Second Manufacturing Paradigm (SMP). These enabling technologies are all characterized by their use of digital, as opposed to analogue, data elements. The first of these technologies to appear was Numerically Controlled (NC) machine tools. These were followed by Computer Aided Design (CAD), and Computer Aided Manufacturing (CAM) linking CAD to NC machine tools. With the aid of these basic building blocks and some new thinking about manufacturing, such as Group Technology, Synchronous Manufacturing, and Concurrent Engineering, we soon saw Flexible Manufacturing Systems (FMS), manufacturing in "batch-of-one" and the realization of a new manufacturing paradigm, the SMP.

This paradigm is characterized by complexity of product, flexibility of process, speed to market and quality levels never realized in mass production. When properly applied, the SMP improves the manufacturing enterprise on all three dimensions of competitiveness, functionality (including quality), cost, and speed-to-market, not by percentage points but by factors of three to five. The next 20 to 50 years will be spent in fully developing the SMP and applying some of its elements to those industrial sectors where mass production remains the paradigm of choice. Continuous process industries will also reflect changes due to the SMP. The new paradigm calls for increasing technology and capital intensive enterprises where these two factors will account for an even greater percentage of economic growth than the current 75 percent [1].

We shall consider the vision of the future of these activities in turn.

III. A VISION OF MANUFACTURING IN A DIGITAL ENVIRONMENT

A. High-Value-Added Discrete Parts

The Second Manufacturing Paradigm will become dominant in all products of considerable complexity and high-value. Aerospace, automotive and weapons systems are typical of this class of product. Most new products now in the conceptual or design stage will also be in this class. Even those products that are to be mass produced will be designed and produced in enterprises organized according to SMP technologies.

We can see already isolated plants where the SMP already is almost fully implemented: Allen-Bradley (Minnesota), Motorola (Florida), Deere (Iowa), General Dynamics (Texas), and IBM (Kentucky & Texas) are examples. In these plants, the enterprise is almost fully integrated. All departments from design, engineering, and manufacturing, to purchasing, sales and marketing share common data bases. In addition, some form of concurrent design is used, and almost the entire fabrication is carried out by flexible manufacturing cells with scheduling done by sophisticated Material Resource Planning (MRP) systems making extensive use of modeling.

By mid 21st century, we will see tens of thousands of plants using such Computer Integrated Manufacturing (CIM) rather than the dozens we have now. As our ability to capture and store knowledge increases, the data bases will increasingly become "Knowledge Bases," until Knowledge Based Manufacturing becomes a fact. Already we see research attempts to capture "Design Intent" and "Expert Systems" to aid in diagnosis and maintenance beginning to reach the factory floor.

Early in the century, we will also see the technology penetrate down to the 350,000 small manufactures. Only in the last few years have these firms had affordable, CIM-capable hardware and software available. The new small plants will be capable of wide product mixes and capable of rapid introduction of a new product to the production line with only software changes. Increasingly the smaller firms will be capable of, and will adopt, a business strategy of Multiple Niche Competition, a concept just now attracting academic interest in our business schools.

Perhaps the principle difference in the digital manufacturing environment in the future will be the ability to do Distributed Enterprise Integrations. In all of the present instances of CIM, each system is operating in a closed proprietary environment. Any linkages with suppliers or customers or between geographically separated sites, are weak and enabled by use of the Initial Graphics Exchange Specification (IGES). IGES is capable of digitally storing and transmitting mechanical drawings, which in the past have been re-entered manually into the CIM systems of the receiver.

In fact, the greatest present lack in enabling technology for the SMP is the lack of a digital manufacturing data structure that will play the same role that the mechanical drawing did in the FMP. This lack is fully recognized by the world-wide manufacturing community and most developed nations have national efforts aimed at producing an International Standards Organization "Standard for the Exchange of Product Model Data (STEP)." The U.S. program is known as "Product Data Exchange using STEP (PDES)." Full functionality of STEP is probably only a decade away and by mid-century will become as ubiquitous as the engineering drawing is today.

In the future, we will see complete manufacturing data freely exchanged across the entire manufacturing community through open, universally accepted, standards, such as STEP. This capability will result in true Distributed Enterprise Integration where small and medium suppliers can contribute in a manner that gains the advantages of total vertical integration without the disadvantages of giantism. Motorola already refers to its culled supplier list as "partners." Design and engineering will certainly be done concurrently involving all elements of the enterprise, including its supplier network.

If we focus our attention on the factory floor we will, in broad perspective, see relatively little change. In the case of mechanical manufacturing, we will still see predominantly metal cutting and forming machines organized into cells connected by material handling systems to form FMS. True there will be improvements in the machines, higher spindle speeds, higher horsepower and better tooling, more accuracy and repeatability, but many of these improvements will be incremental.

Many of the differences from present practice will be hidden, the machine controllers will be "smarter" and standardized. These new controllers will have much better human/machine interfaces, quite possibly two-way voice as well as visual displays, and will increasingly have in-built diagnostic and error recovery capability. The end result will be autonomous machine tools capable of working effectively in cooperation with humans in ways that take the best advantage of the unique capabilities of each. In many cases, manufacturing will be done under what now would be considered clean-room conditions.

The material handling, although similar in concept to today's best practice, will differ in implementation. We will see multi-arm mobile robots with vision, touch, and force sensing which will be able to travel rapidly from place to place, move and position heavy loads, manipulate parts and tools, build and install fixtures, and carry out operations at high speeds on small lot quantities. Scale will vary from many meters to micrometers, and load capacity from many tons to fractions of a gram.

Quality control will be done almost completely by real-time process control, where the process is monitored by a multitude of sensors so that there is no need to inspect the product since it was made by a process known to be in control <u>at the instant that example of the product was produced.</u> Defects in simple product will be at the part per million level and for products as complex as an automobile, the current best performance of 63 customer-discovered defects per hundred vehicles [2] that are discovered by the customers will seem like the "dark ages." The World Class Customer will demand much better performance.

For some 21st century products we will see, however, the growth of new specialized fabrication technologies at sites throughout the industry. The use of high powered lasers for cutting, welding, heat treatment and surface modification will become more common, especially for use with the new high performance materials which will come onto the market. For those hard-to-work materials and to meet the demands for ever closer tolerance and finer finishes, single-point diamond turning, electric discharge machining, abrasive jet machining, and ductile regime grinding will become much more common.

Developing somewhat more slowly will be new "near-net-shape" processes. These attack the fundamental inefficiency of carving complex parts from bulk material. Often the finished part weighs only ten to twenty percent of the initial blank and the rest is reduced to metal chips which are either scrap or must be reprocessed. We see in use now various metal spraying processes that create a product, or a portion of a product, by building-up rather than cutting-away. Widely applied are various compacting processes that start with powder metals or ceramics. Lay-up of composites with pre-impregnated tape is similar in concept and its use is growing rapidity and will become increasingly important.

Rapid prototyping by laser-induced polymerization of plastic is already out of the laboratory. The nano-technology community is currently talking about taking the current semiconductor processing technology to the limit and building IC's atom by atom, rather than layer by layer as is now the practice. New twenty first century products with dimensions of micrometer or nanometer will develop for which only these fabrication technologies will suffice [3].

At this time it is impossible to predict whether near-net-shape technology will become the dominant technology of the SMP or will remain a niche technology. Much will depend on progress in the development of new materials adapted to this "near-net-shape" technology.

Assembly operations are somewhat of a special case. The improved design capabilities will result in products easier to assemble out of larger components made possible by better joining methods. This development will favor the use of robots whose capabilities will be continually enhanced by improvements in their sensory ability. On the other hand the increasingly short runs and wide products mixes will favor human assembly making use of the superior human sensory ability and dexterity. What will develop is increased human/robot cooperation with each doing what they do best.

In any event the production processes will be truly flexible so that the economical-lot-size (ELS) is one, and the variants produced on a single production line will number in the thousands. The concepts of Genichi Taguchi [4], of continuous reduction of variation of product and robust design, will become a way-of-life to a degree that the line itself becomes a dynamic entity, ever improving itself.

B. Mass Production Industries

There will be some industries where the FMP will appear to continue to prevail: for simple products where the batches are huge and the product life-cycle long. Most of these products are those developed during an earlier era, like small arms ammunition, beverage cans, lead pencils and paper clips. For such product the mass production paradigm will continue to show lowest unit cost at acceptable quality. This sector will use the integration and management practices of the SMP for "business" operations and what design must be done.

However, the "hard-automation" typical of mass production also will show increased emphasis on flexibility typical of SMP.

However, in this case, the production line will not be made flexible by the use of changes in the driving software, typical of the SMP, but rather by the use of "metamorphic" or "quick-change" machine hardware. We have seen this technology explored by the Japanese program on "Flexible Manufacturing Complex with Laser" in the late 1970's [5] where they designed an internal combustion engine line capable of being changed over from one to two, to three to four, cylinder product in less than one shift. Already, in the same vein Toyota now changes auto body dies in one hour as compared to the usual two weeks of older practice. These types of technologies will continue to improve and find wider application.

This new technology will give the mass production industries greatly increased "surge" capacity, since with proper advanced planning, in the case of national defense emergencies, entire industries could convert to defense production in days or weeks. No longer, as in World War II, would whole assembly lines have to be scrapped and slowly rebuilt before the demand could be met.

C. Continuous Process Industries

The continuous process industries have been slower to move into the digital environment because of being well along towards full automation by the use of analogue techniques at the birth of the SPM. In 1989 according to figures from the National Center for Manufacturing Science, this industry spent less than one tenth as much for digital equipment as did the discrete batch industry. We can, however, see many of the new trends being adopted. The steel producing mini-mill, and increased emphasis on high-value specialty chemicals and pharmaceuticals, as opposed to bulk chemicals, is a reflection of the SMP characteristic smaller size and flexibility.

The management and integration techniques of SPM will be adopted. The greatest change will be in the "design" phase, where scale-up will be performed by computer modeling and emulation rather than by a succession of ever larger pilot plants. To further this effort, the chemical industry has already formed a consortium to standardize process data in a move to coordinate with PDES. By early in the decade the standard will be in place and in wide use.

This improved process modeling capability will also allow the industry to move closer to the goal of "Total Processing;" that is, the elimination of the waste stream by having no waste, only by-product. The meat packing industry has almost reached this goal. The pressures to protect the environment will force this effort to accelerate [6]. The materials industry will continue to produce new materials and improve old ones, especially in the field of plastics and composites.

D. Staffing and Management

Although we have considered the hardware and software of the factory of the twenty-first century, and only by implication the human staff and its management, it cannot be overlooked that the demands of the SMP will result in changes in staffing patterns and human resources management equally profound. The development of the FMP resulted in a continual "deskilling" of humans on the factory floor. With the SMP, this process is being reversed. The concept of Fredrick Taylor [7] that there exist one best way of doing every task and management must enforce this procedure will fall by the wayside. Management will increasingly become aware that, in the words of Nancy L. Badore of Ford Motors, that involvement and empowerment of the employee is the modern paradigm for management success [8]. The concept of continuous process improvement, propounded by Genichi Taguchi and now widely adopted, assumes on the factory floor a degree of analytical ability and a knowledge of the fundamentals of the processes in use that has been missing from manufacturing since before 1855. These mental, intellectual skills will replace almost entirely the manual effort of the FMP. If the job can be done by "brute strength and ignorance" it will be done by a robot. Since 1900 the percentage of the work force directly engaged in production has decreased and those engaged in management and administration, and scientific technical has increased [9]. There is little reason to believe this trend will not continue until the latter two groups almost completely dominate the manufacturing workforce.

At the higher levels of a much flatter and more open management hierarchy, profound changes in "lifestyle" will occur. Increasingly, design and planning will be done by teams. These teams will use ever more sophisticated management tools and performance metrics [10]. To be successful, such teams will require of their members much greater interpersonal and communication skills and a shift away from individual to team responsibility and reward. Our university engineering and business schools will have adjusted their curricula to provide their graduates with these new necessary skills. Increased emphasis on design technology will be common in engineering courses both undergraduate and graduate [11]. All the necessary changes can be brought about only partly by vocational training and professional education. Improved in education, math, and science starting before entering the work force will be necessary, along with cultural changes.

The introduction of the FMP brought about profound societal changes, such as the rise of the industrial proletariate, in all nations. There is every reason to believe the changes in society brought about will be as great.

IV. CONCLUSIONS

The next 20 to 50 years will see profound changes in manufacturing as the Second Manufacturing Paradigm matures. Those firms that stay at the forefront will prosper as World-Class Producers. Those firms that do not adapt will fail.

There is much work at the research and development level to be done in the United States by industry, government and academia if U.S. Industry is to regain its world class status. Perhaps, most importantly, the industrial culture must change, Taylorism must vanish, internal barriers must fall and management must be as flexible as the means of production. An era will have passed.

REFERENCES

1. M.J. Boskin and L.J. Lau, "Capital Formation and Economic Growth," Technology & Economics, National Academy Press, Washington, DC (1991).

2. J.D. Power & Associates Initial Quality Survey, quoted in Manufacturing Competitiveness Frontiers, February 1991.

3. K. Gabriel, J. Jarvis, and W. Trimmer, "Small Machines, Large Opportunities," Report on Workshop on Microelectromechanical Systems, January 1988, National Science Foundation, Washington, DC (1989).

4. G. Taguchi and D. Clausing, "Robust Quality," Harvard Business Review, January-February 1990.

5. Ministry of Industrial Science and Technology, "National R&D Project, Flexible Manufacturing System Complex with Laser," Japan (1980).

6. G. Heaton, R. Reppetto and R. Sobine, "Transforming Technology: An Agenda for Environmentally Sustainable Growth in the 21st Century," World Resources Institute, Washington, DC (1991).

7. J. Gies, "Automating the Worker," American Heritage of Invention and Technology, American Heritage, New York, NY, Winter 1991.

8. N.L. Badore, "Involvement and Empowerment: The Modern Paradigm for Management Success," Symposium on Foundations of World-Class Manufacturing Systems, National Academy of Engineering, June 19, 1991.

9. D.A. Swyt, "The Workforce of U.S. Manufacturing in the Post-Industrial Era," Journal of Technological Forecasting and Social Change, Vol. 24 (1988).

10. K.R. Baudin, "Manufacturing System Analysis," Youdon Press, Englewood Cliffs, NY (1990).

11. Manufacturing Studies Board, "Improving Engineering Design," National Academy Press, Washington, DC (1991).

INDEX

Q

U

U. S. Government, concurrent engineering and, 56–59
Uncertainty, dimensional measurements and, 127
Unit of measure, dimensional measurements, 118–119, 129
error of, 131
summary and analysis, 155–157

V

Vacuum-wavelength error, error-budget example, 147
Validation procedures
concurrent engineering and, 67–77
STEP program and, 75
NIST national testbed, 87
Validation Testing System, NIST national testbed, 88–90
Value judgment (VJ) module
hierarchical control systems, 201–202
intelligent system architectures, 243–245
Value state-variable map overlays, 245
Value systems, intelligent machines, 199
Verification procedures, concurrent engineering and, 75
VHDL standards, concurrent engineering and, 66
Visible light of known wavelength, 119

W

Wavelength in vacuum, errors in, 132–133
Whitney, Eli, 18
Working groups, concurrent engineering and, 60

Workstation-level decision-making, 259–260
World coordinates, intelligent system architectures, 222
World model, intelligent system architectures, 198–199, 216–229
entities, 225–227
database hierarchy, 227–228
events, 228–229
knowledge representation, 218
maps, 219
egospheres and, 221–225
map-entity relationships, 226–227
overlays, 219–220
pixel frames, 220–221
resolution, 221
sensory processing and, 235–237
space, 219
WM and KD modules, 216–218
World model (WM) module
hierarchical levels, 204
control systems, 201–202
intelligent system architectures, 216–218
sensory processing and, 231

Y

Youden, W. J., 20–21

Z

Z displacement errors, 174–178
Zero, dimensional measurements, 129
error of, 131
summary and analysis, 154–155